Global Issues Series

General Editor: **Jim Whitman**

This exciting new series encompasses three principal themes: the interaction of human and natural systems; cooperation and conflict; and the enactment of values. The series as a whole places an emphasis on the examination of complex systems and causal relations in political decisionmaking; problems of knowledge; authority, control and accountability in issues of scale; and the reconciliation of conflicting values and competing claims. Throughout the series the concentration is on an integration of existing disciplines towards the clarification of political possibility as well as impending crises.

Titles include:

Malcolm Dando
PREVENTING BIOLOGICAL WARFARE
The Failure of American Leadership

Brendan Gleeson and Nicholas Low (*editors*)
GOVERNING FOR THE ENVIRONMENT
Global Problems, Ethics and Democracy

Roger Jeffery and Bhaskar Vira (*editors*)
CONFLICT AND COOPERATION IN PARTICIPATORY NATURAL RESOURCE
MANAGEMENT

Ho-Won Jeong (*editor*)
GLOBAL ENVIRONMENTAL POLICIES
Institutions and Procedures
APPROACHES TO PEACEBUILDING

W. Andy Knight
A CHANGING UNITED NATIONS
Multilateral Evolution and the Quest for Global Governance

W. Andy Knight (*editor*)
ADAPTING THE UNITED NATIONS TO A POSTMODERN ERA
Lessons Learned

Kelley Lee (*editor*)
HEALTH IMPACTS OF GLOBALIZATION
Towards Global Governance

Graham S. Pearson
THE UNSCOM SAGA
Chemical and Biological Weapons Non-Proliferation

Andrew T. Price-Smith (*editor*)
PLAGUES AND POLITICS
Infectious Disease and International Policy

Michael Pugh (*editor*)
REGENERATION OF WAR-TORN SOCIETIES

Bhaskar Vira and Roger Jeffery (*editors*)
ANALYTICAL ISSUES IN PARTICIPATORY NATURAL RESOURCE MANAGEMENT

Simon M. Whitby
BIOLOGICAL WARFARE AGAINST CROPS

Global Issues Series
Series Standing Order ISBN 0–333–79483–4
(*outside North America only*)

You can receive future titles in this series as they are published by placing a standing order. Please contact your bookseller or, in case of difficulty, write to us at the address below with your name and address, the title of the series and the ISBN quoted above.

Customer Services Department, Macmillan Distribution Ltd, Houndmills, Basingstoke, Hampshire RG21 6XS, England

Health Impacts of Globalization

Towards Global Governance

Edited by

Kelley Lee
*Senior Lecturer in Global Health Policy
and Co-Director of the Centre on Global Change and Health
London School of Hygiene and Tropical Medicine*

First published 2003 by
PALGRAVE MACMILLAN
Houndmills, Basingstoke, Hampshire RG21 6XS and
175 Fifth Avenue, New York, N.Y. 10010
Companies and representatives throughout the world

PALGRAVE MACMILLAN is the global academic imprint of the Palgrave
Macmillan division of St. Martin's Press, LLC and of Palgrave Macmillan Ltd.
Macmillan® is a registered trademark in the United States, United Kingdom
and other countries. Palgrave is a registered trademark in the European
Union and other countries.

ISBN 0–333–80254–3

This book is printed on paper suitable for recycling and made from fully
managed and sustained forest sources.

A catalogue record for this book is available from the British Library.

Library of Congress Cataloging-in-Publication Data

Health impacts of globalization: towards global governance /
edited by Kelley Lee.
    p. cm. – (Global issues)
   Includes bibliographical references and index.
   ISBN 0–333–80254–3
     1. Globalization–Health aspects. 2. World health. I. Lee, Kelley,
    1962- II. Series.

   RA441 .H435 2002
   362.1–dc21   2002074845

10 9 8 7 6 5 4 3 2
12 11 10 09 08 07 06 05

Printed and bound in Great Britain by
Antony Rowe Ltd, Chippenham and Eastbourne

For Andrew, Jennifer and Alexander

Contents

List of Tables, Figures and Boxes

Tables

Figures

Boxes

Preface

This book started as a modest, and rather selfish, attempt to make sense of seemingly idiosyncratic career choices. Armed with broad training in international relations (IR), I embarked on a two-year stint at the London School of Hygiene and Tropical Medicine in 1992 to analyse the role of the United Nations in the health field. Health was a Cinderella topic in terms of the study of international organization, but almost immediately I recognized the substantive contribution that IR could bring to health research and policy. At the same time, the puzzling neglect of health issues by the IR community came to my attention.

Ten years later, and with the advantage of hindsight, the rather strange decision to join an internationally renowned school of public health now makes perfect sense. The 1990s was a period of wide-ranging change in the health sector, with more and more issues bumping up against the boundaries of traditional health research and policy. Of particular note is the recognition that so many determinants of health could not be, and perhaps never were, confinable to national boundaries. The field of global health is rapidly coming into its own, supported by a need to grapple with issues that cross familiar geographical, disciplinary and sectoral boundaries. Equally important has been the significant attention to health by a wider range of prominent individuals and institutional actors. Health has never been higher on the agenda of 'high politics', billed as either part of the 'new' risks to national security, or as an opportunity to renew commitment to international co-operation in a post-Cold War order. Amid these rapid changes, it has never been a better time to bridge the disciplinary divide between the disciplines of IR/politics and health.

The chapters of this volume represent this bringing together of these two broad fields, drawing on the growing critical mass of writers who have challenged the boundaries of their respective disciplines. I have been fortunate to work with almost all of the authors in this volume in varying capacities, and have benefited immensely from their diverse experiences and insights.

Acknowledgements

A large number of people have helped to shape the structure and ideas of this book, and its accompanying volume, in varying ways to which I am immensely grateful. My warm appreciation goes to John Wyn Owen, Secretary of The Nuffield Trust whose personal interest in the subject of globalization and health predates most in the field today. Generous funding by the Trust itself provided the foundations upon which so much of the attention to global health has been built in the UK and abroad. He continues to be a dedicated messenger for spreading the word among those who have the power to shape globalization into a more progressive and humane force. Also working with me at the Trust is Graham Lister who has been a supportive and energetic colleague in creating and maintaining the UK Partnership for Global Health. This is a unique endeavour that brings together a wide range of interest groups to find ways forward for tackling the shared health challenges of globalization.

My thanks also go to staff at the World Health Organization with whom I have worked in close relationship to build international recognition of the new challenges of globalization. Nick Drager, Coordinator of the Programme on Globalization, Trade and Cross Sectoral Issues has been a continual source of encouragement in this work. Over the years, he has provided valuable links to the developing world which have enriched my work immensely. The ideas and comments of other WHO personnel within the globalization programme, notably Robert Beaglehole, Debra Lipson and David Woodward, have prompted me to think through the practical relevance of this work more carefully. The varied members of the Scientific Resource Group on Globalization, Trade and Health also provided much needed reality checks when I tended to go off on my academic musings. Douglas Bettcher and Derek Yach of the Tobacco Free Initiative also provided key insights into some of the chapters in this book, which I appreciate.

At the London School of Hygiene and Tropical Medicine, I have had informative inputs about conceptual and empirical issues related to this volume with Karen Bissell, David Bradley, Anna Gilmore, Sari Kovats, Dave Leon, Martin McKee, Tony McMichael, Peter Smith, Carolyn Stephens and Paul Wilkinson. Nick Black has been particularly supportive in giving the subject of globalization a higher profile within the School, facilitated by former Dean Harrison Spencer and our current Dean Andy Haines.

At Yale University, I am indebted to Ilona Kickbusch for inviting me to unleash my early ideas on her public health students, and to Kent Buse for his continued friendship and intellectual energy. I also spent an enjoyable two weeks at Yale in summer 1999 attending the Academic Council on the United Nations System (ACUNS) Summer Workshop on global governance. The experience energized my desire to take this work forward.

In the final stages of pulling this collection together, I am indebted to my colleague Preeti Patel who took on the monumental task of shepherding this book, and its accompanying volume, to the publisher. Without her patient support and attention to detail, the two books would still only be an elusive dream yet to be achieved. Barbara Cannito provided additional research assistance in ferreting out elusive reference material. Tamsin Kelk undertook copy editing of the chapters with her usual efficiency. Phillip Raponi, Lucy Paul, Linda Amarfio and Nicola Lord kindly gave administrative support whenever I stumbled past their offices in search of assistance.

Finally, I am grateful as always to my husband Andrew for his kindness and patience over the years as this work has slowly developed. It is only with his support that I have been able to run off to so many far flung places at times when the challenges of a young family seem insurmountable.

The publishers would like to thank *New Political Economy* for granting permission to reprint Kelley Lee and Antony Zwi's chapter originally published in vol. 1 no. 3, 1996, *Third World Quarterly* for allowing us to reprint Caroline Thomas's chapter, which appears in a special edition, vol. 23(2), published in April 2002 and *Global Governance* for Kelley Lee and Richard Dodgson's chapter published in vol. 6, no. 2, 2000.

Notes on the Contributors

Dennis Altman is Professor of Politics at LaTrobe University, Melbourne, and author of ten books, most recently *Global Sex* (2001), which was written in part with the support of a Macarthur Foundation Research and Writing Fellowship. His other books relevant to AIDS are *Power and Community* (1994) and *AIDS and the New Puritanism* (1986). He was a founding member of the AIDS Society of Asia and the Pacific and the International Council of AIDS Service Organizations, and in 2001 co-chaired the Sixth International Congress on AIDS in Asia and the Pacific.

Mickey Chopra worked for four years as a medical officer at Hlabisa district in rural South Africa shortly after completing his medical internship. After completing a M.Sc. (Public Health in Developing Countries) at the London School of Tropical Hygiene and Medicine he joined the staff at the School of Public Health, University of the Western Cape. His main interests are in the fields of health systems research, child health and health programme development.

Jeff Collin is a Lecturer in the Centre on Global Change and Health at the London School of Hygiene and Tropical Medicine. He has completed a project funded by The Nuffield Trust to review public health measures to control transborder health risks in the United Kingdom in the context of globalization. With a primary research interest in tobacco control, he is currently participating in a 14-country study for the US National Institutes of Health entitled 'Globalization, the Tobacco Industry and Policy Influence'.

Richard Dodgson is a research officer in widening participation at Universities for the North East. In this post he has recently completed the first stage of a comparative project that examines the widening access policies on a global scale. Prior to taking up his current post, he received his doctorate at the University of Newcastle upon Tyne for research published in the areas of globalization, international political economy and the global politics of health/population control.

David P. Fidler is Professor of Law at Indiana University School of Law, Bloomington, Indiana, USA. He is a leading scholar in the area of international law and public health. His publications include *International Law and Infectious Diseases* (1999) and *International Law and Public Health: Materials on and Analysis of Global Health Jurisprudence* (2000). His articles on legal aspects of antimicrobial resistance have appeared in *Emerging Infectious Diseases* and *Microbes and Infection*. He has acted as an international legal consultant on antimicrobial resistance to the World Health Organization,

US Centers for Disease Control and Prevention, and the American Academy of Microbiology.

Ilona Kickbusch is Head of the Division of Global Health at Yale University School of Medicine, in the Department of Epidemiology and Public Health. Her major research interests are in global health policy and governance, partnerships for health development, healthy communities and social determinants of health. Present research projects include health literacy, international health promotion development and health and security. Dr Kickbusch joined Yale after a long career with the World Health Organization where she initiated the Ottawa Charter for Health Promotion and headed a range of innovative programmes such as Healthy Cities and Health Promoting Schools. As director of Communication at WHO/HQ in Geneva she oversaw the planning for World Health Days and the health pavilion at the World EXPO 2000 in Hanover.

Meri Koivusalo completed her medical degree and Ph.D. in epidemiology at Helsinki University and received a M.Sc. in Environmental Health Policy from the London School of Hygiene and Tropical Medicine. She is currently a Senior Research Fellow in the Globalism and Social Policy Programme at the Finnish National Research and Development Centre for Welfare and Health (STAKES) in Helsinki, Finland. She continues to work as an adviser on international health policy to the Finnish Ministry of Social Affairs and Health and the Ministry of Foreign Affairs. She has published widely and co-authored *Making a Healthy World* (1997). Her current work deals with international, regional and national health policy implications of multilateral trade agreements.

Kelley Lee is Senior Lecturer in Global Health Policy and Co-Director of the Centre on Global Change and Health at the London School of Hygiene and Tropical Medicine. She chairs the WHO Scientific Resource Group on Globalization, Trade and Health, and is a founding member of the UK Partnership for Global Health. She is involved in a number of projects analysing the impacts of globalization on public health with particular interests in infectious diseases, tobacco control and global governance. She is the author of *A Historical Dictionary of the World Health Organization* (1999), *Health Policy in a Globalising World* edited with Kent Buse and Suzanne Fustukian (2002) and *Globalization and Health: an Introduction* (forthcoming).

Preeti Patel is a Research Fellow in the Centre on Global Change and Health in the Department of Public Health and Policy at the London School of Hygiene and Tropical Medicine. Her current research interests include: globalization and the political economy of tobacco control, politics of AIDS in Africa, international peacekeeping and multiparty politics in Kenya.

David Sanders has since 1993 been Director and Professor of a new Public Health Programme at the University of the Western Cape, and since 2000 Deputy Dean of the Faculty of Community and Health Sciences. The Public Health Programme (PHP) provides practice-oriented education and undertakes research in public health and primary health care. The PHP launched South Africa's first Master of Public Health (MPH) programme and is now nationally recognized for its service development work in the areas of health information, community nutrition and school health. Between 1980 and 1992 he lived and worked in Zimbabwe where he was Medical Adviser to OXFAM (UK) in setting up rural health programmes and was actively involved in restructuring of that country's health service. He was at the same time a member of the academic staff of University of Zimbabwe Medical School, first in the Department of Paediatrics and later the Department of Community Medicine. Since returning to South Africa he has been actively involved in the health policy process. David Sanders is author of *The Struggle for Health: Medicine and the Politics of Underdevelopment* and co-author of *Questioning the Solution: the Politics of Primary Health Care and Child Survival* and of numerous articles on the political economy of health, structural adjustment, child nutrition and health personnel education.

Roy Smith is Co-Director of the Centre for Asia-Pacific Studies, Course Leader for B.A. (Hons) International Relations and Senior Lecturer, in the Department of International Studies at Nottingham Trent University. His teaching areas include: International Political Economy, International Relations of the Environment, the Political Economy of Australasia and the Pacific Islands. His has research expertise on Small Island States, health issues, environmental issues and cyber-politics.

Caroline Thomas is Professor of Global Politics at Southampton University. She has a longstanding interest in South–North issues and has published widely in this field. In recent years her research and publications have focused on the impact of globalization on the South. Her most recent book is *Global Governance, Development and Human Security* (2000).

Anthony Zwi is head of the School of Public Health and Community Medicine at the University of New South Wales, Australia. He was born and grew up in South Africa where he undertook his medical degree and postgraduate diploma in occupational health as well as tropical medicine. He subsequently trained in epidemiology, public health and international health at the London School of Hygiene and Tropical Medicine, and in the United Kingdom National Health Service. He headed the Health Policy Unit at the London School of Hygiene and Tropical Medicine from 1997 to 2000 and has actively promoted the study of public health and health systems in humanitarian crises. He has longstanding interest in how conflict impacts upon health and health systems and has promoted the establishment of a network to adapt and transfer analytic tools and response strategies between countries emerging from major periods of conflict. He is committed to promoting mechanisms whereby affected communities can make their voices heard, policymakers and the policies they promote can be made more accountable, and academic–NGO partnerships can be promoted to ensure that lessons are learned and mistakes not repeated in humanitarian and development contexts.

List of Abbreviations

AIDS	Acquired Immune Deficiency Syndrome
AME	antimicrobial effectiveness
AMR	antimicrobial resistance
APUA	Alliance for the Prudent Use of Antibiotics
ARV	anti-retroviral
AZT	zidovudine
BAT	British American Tobacco
BSE	bovine spongiform encephalopathy
CAC	Codex Alimentarius Commission
CBO	community-based organization
CECCM	Confederation of European Community Cigarette Manufacturers
CIA	Central Intelligence Agency
CIL	customary international law
CIS	Commonwealth of Independent States
CJD	Creutzfeldt-Jakob disease
CPT	Consumer Project on Technology
CRO	contract research organization
CSIS	Center for Strategic and International Studies
DALY	disability adjusted life year
DEFRA	Department of Environment, Food and Rural Affairs (since June 2001) (UK)
DOH	Department of Health (UK)
DOTS	directly observed therapy, short-course
EC	European Commission
ECOSOC	United Nations Economic and Social Council
EEC	European Economic Community
ESAP	Economic Structural Adjustment Programme
ETS	environmental tobacco smoke
EU	European Union
FAO	Food and Agriculture Organization
FCTC	Framework Convention on Tobacco Control
FDA	Food and Drug Administration (US)
G8	Group of Eight countries (Canada, France, Germany, Italy, Japan, Russia, US and UK)
G77	Group of 77 countries
GATS	General Agreement on Trade in Services
GATT	General Agreement on Tariffs and Trade
GDP	Gross Domestic Product

GHG	global health governance
GIPA	greater involvement of people living with AIDS
GMO	genetically-modified organism
GNP	Gross National Product
GPA	Global Programme on AIDS
GPE	global political economy
GSK	Glaxo Smithkline
GSP	Generalized System of Preferences
HFA	Health for All
HIV	human immunodeficiency virus
IARC	International Agency for Research on Cancer
IAVI	International AIDS Vaccine Initiative
IEL	international environmental law
IFC	International Finance Corporation
IHR	International Health Regulations
ILO	International Labour Organization
IMF	International Monetary Fund
IMR	infant mortality rate
INFOTAB	International Tobacco Information Centre
IPE	international political economy
IR	international relations
ISO	International Standards Organization
ITC	Indian Tobacco Company
IVDU	intravenous drug users
LRTAP	Long-Range Transboundary Air Pollution
MAFF	Ministry of Agriculture, Fisheries and Food (UK)
MBM	meat and bone meal
MCH	maternal and child health
MDRTB	multi-drug resistant tuberculosis
MNC	multinational company/corporation
MSF	Médicins Sans Frontières
MTA	multilateral trade agreement
NAFTA	North American Free Trade Agreement
NCD	noncommunicable disease
NGO	nongovernmental organization
ODA	official development assistance
OECD	Organization for Economic Cooperation and Development
OIHP	Office International d'Hygiène Publique
OPD	outpatient department
ORT	oral rehydration therapy
PHC	primary health care
PLWHA	people living with HIV and AIDS
ProMED	Program for Monitoring Emerging Diseases
PrP	'prion' protein

R&D	research & development
RHC	rural health centre
SADC	Southern African Development Community
SAP	Structural Adjustment Programme
SEAC	Spongiform Encephalopathy Advisory Committee
SPS	Sanitary and Phytosanitary Measures
SSA	sub-Saharan Africa
STD	sexually transmitted disease
STI	sexually transmitted infection
TB	tuberculosis
TRIPS	Trade-Related Intellectual Property Rights
TSE	transmissible spongiform encephalopathies
TTC	transnational tobacco company
UK	United Kingdom
UKIB	UK International Brand
UN	United Nations
UNAIDS	United Nations Joint and Co-sponsored Programme on AIDS
UNDP	United Nations Development Programme
UNFPA	United Nations Population Fund
UNECE	United Nations Economic Commission for Europe
UNESCO	United Nations Educational, Scientific and Cultural Organization
UNGASS	United Nations General Assembly Special Session on AIDS
UNICEF	United Nations Children's Fund
US	United States
USAID	United States Agency for International Development
USIB	US International Brand
USTR	US Trade Representative
vCJD	variant Creutzfeldt-Jakob disease
WHA	World Health Assembly
WHO	World Health Organization
WTO	World Trade Organization

1
Introduction
Kelley Lee

1.1 Bridging the divide: globalization and health

Like so many aspects of our lives since autumn 2001, the writing of this introduction is dominated by thoughts of the terrorist attacks on the World Trade Center and the Pentagon, and their aftermath. Undoubtedly, there will continue to be much to reflect on as the full consequences of the events are played out around the world. However, one of the immediate messages brought home by them is the stark realization that globalization is a double-edged sword. For those of us fortunate enough to be able to advantage ourselves of the many benefits of an increasingly interconnected world, the events of 11 September shook our basic confidence in 'the system'. By and large shielded from the blunter end of globalizing processes, the 'winners' of globalization have come to enjoy such features as 24-hour financial trading, worldwide travel to far flung places for work and leisure, a seemingly infinite array of information sources via the Internet, and a huge range of consumer goods and services from burgeoning international trade. What is now also evident is that global financial systems can be used to launder huge sums of money from questionable sources for dubious purposes; mass transportation across international borders can enable terrorist organizations to move about often unnoticed; and global communication systems can be used to create well-organized social networks across the world with good or bad intent.

There are many direct and indirect health consequences of the terrorist attacks. The deaths and injuries, mostly deaths, directly resulting from the hijackings are the most immediate. Almost as dramatic at causing public panic is the prospect of biological and chemical weapons being used by terrorists on civilian populations. The mysterious cases of Anthrax diagnosed in different parts of the US and the prospect of smallpox, plague and other infectious agents being spread intentionally have raised concerns about the capacity of governments to effectively prevent such attacks from nonstate actors. The so-called 'War against Terrorism' initiated by the US and its allies, beginning with military action against Afghanistan, is leading to further death and injury among military personnel and civilians. The humanitarian crisis prompted by these actions is also important to the tallying up of health impacts.

It is with this backdrop that the final stages of compiling this book, and its accompanying volume, were completed. There is a danger, of course, of overstating the effects of the above events on international relations. While the world looks a very different place through the lens of large-scale and transborder terrorism, clearly most of the world's woes predate these events. Furthermore, it would be ill advised to reduce the health consequences of globalization only to the risks that the attacks have poignantly brought into focus – terrorist violence, biological and chemical weapons, and armed conflict. To develop effective responses to such risks, we need to go beyond traditionally defined threats to national security. Why so much anti-western feeling in a world supposedly becoming more interconnected as a result of globalization? What concerns lie at the heart of the mass demonstrations at major international meetings in recent years? And, most difficult perhaps, why is the emerging global order such a fertile ground for feelings of alienation and marginalization among so many people? Ignorance, envy, conservatism or plain evil – none are satisfactory explanations. Without learning broader lessons, including how globalization must be more inclusive and equitable, we will not tackle the true roots of so many global health risks.

In seeking to contribute to a fuller understanding of the risks and opportunities of globalization, the origins of this book are twofold. First, there remains limited analysis of the global dimensions of health within the mainstream globalization literature as a whole, including the fields of international relations (IR) and politics. With a few notable exceptions, there are no concerted forays into health-related issues by IR scholars comparable, for example, to the environment, human rights and gender (Baylis and Smith, 1997). There are several reasons for this. Historically, IR is strongly based on the applied concerns of diplomacy and foreign policy, resulting in a preoccupation with peace and security. Health, in contrast, is seen as a topic appropriate for the table of 'low politics', a social welfare issue that is primarily dealt with at the national level by the governments of states. Perhaps a further factor is the rough equating of health research with the study and practice of medicine, a field that is perceived as largely 'apolitical' given its scientific and technical content. The subfield of public health, for example, is traditionally associated with such practical matters as immunization and sewers. Cast in these terms, health fails to even register on the radars of mainstream globalization scholars.

There has been occasional lament of this neglect of health in IR (Thomas, 1989; Murphy, 2000), and inklings of a shift in the tide as certain health issues (such as HIV/AIDS) catch the imagination of individual scholars (Gordenker *et al.*, 1995). Since the end of the Cold War, the recasting of national security issues in terms of 'new security threats', notably population movements, environmental degradation, organized crime and infectious diseases, have given health an entrée into studies of foreign policy (Institute of Medicine, 1997). The study of health from the broader field of politics has also borne greater fruit. With an established appreciation of the power relations intrinsic to, for example, the allocation of health resources, differences in health status, organization of health care services and distribution of health development aid, there have been wide-ranging studies of the politics of health. Broadly speaking, however, most of this literature has focused on specific countries or, at the most, the comparative national level. Again there are important exceptions (Doyal, 1979), which, while not explicitly focused on globalization, can form the foundation upon which the study of health from a global perspective can usefully build.

The second impetus for this book has been the limited engagement, until relatively recently, by the health community with the diverse and rich literature on globalization. Health care systems are traditionally structured, regulated and delivered at the national level. Health research is correspondingly focused on the state, sub-state and inter-state levels. There is rapidly growing interest in 'global health', although particular issue areas such as the emerging global economy and infectious diseases receive much of the attention (Lee, 2000). Other works, while claiming to address global health concerns, do not go much beyond comparative analysis of national health systems (Altenstetter and Bjorkman, 1997; Coulter and Ham, 2000). The apparent reluctance to directly engage in globalization issues is due, in part, to a degree of scepticism about how distinct the field really is. More importantly, however, is the sheer volume and often technical complexity of many of the writings on globalization. This can be off-putting to the uninitiated, especially when ungrounded in empirical analysis. The importance given to so-called 'evidence-based medicine' does not make the theoretical nature of much globalization scholarship especially appealing. Despite these challenges, the steady growth of global health-related initiatives in major public health institutions, special issues of mainstream health and medical journals, and meetings and conferences on the subject of globalization and health attests to the importance now given to the subject. There is a clear desire for a fuller understanding of the links between globalization and health, beginning with the development of relevant conceptual frameworks and rigorous empirical analysis (Woodward *et al.*, 2001).

Hence, this book represents a sharing of knowledge between the fields of IR/politics and health in order to understand better the impacts of globalization on human health. The disciplinary backgrounds of the contributors to this volume are worth noting because they embody the need to approach the subject of globalization and health from different angles. Dennis Altman, Jeff Collin, Richard Dodgson, Ilona Kickbusch, Kelley Lee, Preeti Patel, Roy Smith and Caroline Thomas hold doctorate degrees in either IR or politics, and come to the health field with varied knowledge and experience. While the balance of their careers is academically based, many have served as advisers to national governments, nongovernmental organizations (NGOs) and international organizations concerned with health including the World Health Organization (WHO), United Nations Children's Fund (UNICEF), United Nations Development Programme (UNDP) and World Bank. Most have experience of research in both high- and low-income countries. In addition, David P. Fidler is an international lawyer with additional training in IR. Ilona Kickbusch's long career in WHO, including as Director of Health Promotion, has invaluably informed her scholarly work. The remaining authors, Mickey Chopra, Meri Koivusalo, David Sanders and Anthony Zwi are medically qualified. They too have gone outside familiar disciplinary boxes, driven by searching questions about the broader determinants of health. Again, their work draws from strong research and policy experience that has taken them to many parts of the world.

In bringing this interdisciplinary group of researchers together, this collection of case studies provides much needed empirical analysis of the impacts of globalization on health. Scholarly and policy debates about such impacts have become increasingly polarized, ranging from liberal-based optimism that globalization is mostly good for our health (Feachem, 2001) to sweeping condemnation that it offers few redeeming features (Baum, 2001). The worrying feature of such debates is that they are too often based on conceptual fuzziness, ideological preconceptions or vested interests, rather than attempts to document, measure and assess the relative balance of positive and negative effects that globalization brings to different individuals and population groups. While there are, of course, clear limitations to empiricism, eschewing the need for empirical evidence leads both sides to a 'dialogue of the deaf' (Lee, 2001). In this sense, the chapters in this book offer ideas for taking forward the current stalemate in policy perspectives through a concerted research agenda on globalization and health.

1.2   Globalization and health: finding a conceptual starting point

Before proceeding farther along this analytical path, the challenge of defining globalization, and hence the parameters of this book, lie before us. The existing literature on globalization is abundant and fast growing, offering a virtual mountain of publishing across many different fields and a broad range of perspectives. A cursory glance at selected works soon reveals a clear difference of opinion as to what precisely is globalization. For economists, globalization is an emerging global economy (Dicken, 1998); for international lawyers, it is possible changes to the legal status of states and their citizens (Rubenstein, 2000); for computer scientists and information specialists, it is the information superhighway (Mohammadi, 1997); and in cultural studies, it is the trend towards a global, strongly western-defined, culture (Jameson and Miyoshi, 1998). As well as recognizing different features of globalization, opinion is widely divided as to the precise impacts of globalization on specific individuals and population groups. Different philosophical and ideological starting points have led to often polarized views on the positive and negative impacts of globalization.

It is, therefore, important to set out here some of the key defining features of globalization and hence the parameters of 'global health' as an emerging field of research and policy. Broadly speaking, globalization can be defined as a set of processes that are changing the nature of how humans interact across three types of boundaries – spatial, temporal and cognitive. A fuller explanation of these dimensions of global change is provided below and more fully elsewhere (Lee, 2002). The changes they are creating are 'global', in the sense that, familiar boundaries separating us as individuals and societies have become increasingly eroded. Most prominent, perhaps, have been the political boundaries of states, with many arguing that globalization is eroding, and even making obsolete, national borders. However, the degree to which states are threatened remains a subject of debate. Definitions of globalization range from their focus on the intensification of crossborder flows across states boundaries, such as international trade, that in many ways reinforce the role of the state (Woodward, 2001), to transborder flows that disregard territorial geography altogether such as the Internet and global environmental change (Scholte, 2000). The latter, it is argued, may undermine state authority. Whatever view one holds about the nature and intensity of the flows taking place, it is recognized that they are impacting on many aspects of our lives. Globalization impacts on a wide range of social spheres including the economic, political, cultural and technological.

Looking at the nature of the changes taking place more closely, the *spatial dimension* of global change concerns how we experience and perceive physical space. On the one hand, there is a growing 'sense of the world as a single place' (Robertson, 1992) due to increased migration, communication, production and exchange, and other shared experiences across the world. In contrast with a world divided into 190-odd sovereign and territorially distinct states, globalization is challenging many factors that define the organization of societies strictly along national borders. The popular and rather idealized image of the 'global village' in which world citizens engage across vast expanses, derives from this perception. On the other hand, there is evidence that globalization is reinforcing existing territorial boundaries or creating new territorial divisions such as new states or regional organizations. Rather than physical space becoming irrelevant, it is being redefined along different parameters. This is perhaps best illustrated by the novel forms of social organization that are emerging. The advent of cyberspace and virtual reality, for example, redefine geography in innovative ways as they are not physically tied to territorial space.

The health implications of global spatial change are wide-ranging. The most direct effect is how patterns of health and disease are altering according to new forms of social organization. Aligned with new patterns of socioeconomic inequity within and across countries, for instance, are corresponding trends in health status such as burdens of disease and birth and death rates. Hence, it is no longer sufficient and even accurate to speak of the population of country X as having a given health status as much international health data continues to do. National aggregates obscure the fact that different population groups, depending on their location in the emerging global order, will have different health needs. If you happen to be among the globally advantaged, it is likely that your health status will be more akin with people of similar advantage in other countries of the world than the globally disadvantaged within your country. In short, new spatial configurations of health and disease are emerging as a consequence of globalization.

As well as changes to patterns of health and disease, global spatial change concerns how the determinants of health are being reorganized along different geographical lines. Foremost is the impact of an increasingly globalized economy, resulting *inter alia* in a global division of labour, supply and demand chains, flows and distribution of capital and terms of trade. The flow of information via global telecommunications is also creating new spatial relationships that impact on health through, for example, marketing and advertising, policy reforms, scientific research and medical education. Similarly, global environmental change is occurring on a scale that makes some of its health impacts potentially worldwide (for example El Nino) and others population specific (for example impact of global warming on coastal communities).

The *temporal dimension* of globalization concerns changes to how we perceive and experience time. In many ways, there is a speeding of the time-frame that many events take place. A notable example is modern telecommunications. The advent of satellite technology, facsimile and the Internet allows messages to be sent and received around the world in microseconds. Because communications underpin so many human interactions, this capacity has speeded such diverse activities as currency trading, news bulletins, business operations of transnational corporations (TNCs) and political elections. Similarly, mass transportation technologies, including high-speed bullet trains and supersonic jets, enable us to travel farther and faster. The faster movement of people, and communication among them, have many implications for health. As people physically move faster about the globe, so too do microbes that accompany them, leading to changes in the pattern of health and disease. However, the ability to communicate faster can also mean enhanced capacity to monitor and report on outbreaks of disease, disseminate guidelines for controlling and treating disease, and coordinate rapid responses when needed.

In other ways, how we interact has been slowed by globalization. As the world that affects our daily lives expands, so too do we need to network with increasing numbers of people, absorb more information sources and choose among products and services available to us. The experience of 'information overload'[1] can occur at an individual and collective level and mean that we need longer to understand tasks and take decisions. Policymakers who need to consult with a larger number and wider range of stakeholders, for instance, may find the process more time consuming. Health professionals who are required to keep up with the latest techniques find it increasingly difficult to do so with so much information readily to hand.

The *cognitive dimension* of globalization concerns changes to how we think about ourselves and the world around us. A variety of thought processes – the creation and exchange of scientific knowledge, ideas, norms, beliefs, values, cultural identities – are being affected by globalizing forces. These forces are varied and include the mass media, advertising agencies, educational institutions, think tanks, scientists, consultancy firms, public relations offices including so-called 'spin doctors', the Internet, international organizations and religious groups. All are seeking to influence how we think and increasingly the scope of their reach is global. This has prompted a rich scholarly literature around cultural globalization.

Once again, the health effects are diverse. Health can benefit from a greater sharing of scientific knowledge and technology across countries. Innovations such as electronic journals, distance-based learning and virtual conferencing open the possibility for much needed health information to be accessed by more people. The definition and international adoption of standards concerned with workers health and safety, human rights, reproductive health, environmental protection and corporate governance can also contribute to improved health. More controversial is global marketing and advertising aimed at increasing consumption of certain foods, drink and tobacco products that are harmful to health. The plethora of health sector reforms promulgated by bilateral and international organizations over recent decades also have the potential to impact profoundly on the performance of health systems around the world.

1.3 The structure of the book

The aim of this book is to bring together interdisciplinary and empirical analysis of the complex and diverse links between globalization and human health. There is an accompanying volume which offers a fuller account of the conceptual framework presented here. At the same time, the case studies stand on their own as a unique collection in the current literature. Importantly, the range of issues covered is intended to be illustrative, rather than representative, of the diverse health impacts of global change. The contributors hope that their work can encourage the further development of a diverse range of research that strengthens empirical evidence of the risks and opportunities of globalization, as well as inform effective policy responses to them.

The organization of this book has been an analytical challenge. By definition, globalization is leading to greater interconnectedness across many issue areas, and the subjects tackled by these chapters invariably rub up against each other in their exploration of health consequences. It is perhaps unavoidable that much of this book is about changing patterns of health and disease. A number of chapters deal with the devastating global HIV/AIDS pandemic, unsurprising given its close association with so many features of the emerging global order. Kelley Lee and Anthony Zwi argue that HIV/AIDS can be more fully understood from the perspective of global political economy in terms of the populations worst affected, the underlying 'risk factors' and the nature of the policy responses so far by the international community. Dennis Altman explores HIV/AIDS as a security issue, arguing that we need to recognize the disease as a serious threat to the integrity of high-burden countries, and through them the national security of all countries. A similar approach is taken by Kelley Lee and Preeti Patel is their discussion of the global dimensions of bovine spongiform encephalopathy (BSE), and its human form variant Creutzfeldt-Jakob disease (vCJD), as intimately tied to the structure of the food, animal feed and biologicals industries. They raise questions about the way our food is produced and the primacy given to economic interests over the exigencies of public health, environment or ethics.

Changing patterns of noncommunicable diseases are also recognized among the impacts of globalization. Jeff Collin explores the cognitive dimensions of globalization through an analysis of internal documents of the tobacco industry documents released to public scrutiny in 1998. The documents reveal the aims and strategies of an industry with long experience of influencing public and official opinion through advertising and sponsorship, manipulation of the scientific process, and policymaking. Collin examines an aspect of globalization that has received limited attention so far, yet his findings raise fundamental questions about the nature and direction of a globalization defined so powerfully by corporate interests. The chapter by Roy Smith comes from primary research carried out among local people on the Marshall Islands. Using focus group interviews, he explores how global economic and social change has influenced attitudes to health and nutrition. Similarly, David Sanders and Mickey Chopra seek to link macrolevel change with microlevel impacts. They are concerned with the adverse health impacts of global economic and social change on local communities in the developing world. They argue, through data collected in southern Africa during the late 1990s, that deteriorating health status in two communities coincides with macroeconomic reforms introduced by the World Bank and other major aid donors.

A second major focus of this book is governance for global health that is primarily, although not exclusively, addressed in Part II. Each of the chapters in the second half of the book provide empirical analysis of a major area of global health, and accompany this with an exploration of the challenges for ensuring effective governance at various levels of policymaking. Kelley Lee and Richard Dodgson present an historical analysis of cholera from the early nineteenth century, linking the spatial and temporal patterns of the disease with broader changes associated with globalization. They argue that particular features of contemporary globalization, namely socioeconomic inequality, rapid movement of human populations, modern transportation technologies and environmental degradation, explain the distinct behaviour of the seventh cholera pandemic dating from the early 1960s. This raises fundamental questions about the existing institutional framework for international co-operation on infectious disease, with its strong emphasis on biomedical understandings of human disease and state-based health systems. The changing nature of disease is also raised by David P. Fidler in this analysis of antimicrobial resistance (AMR). He argues that reliance on non-binding recommendations by WHO to national governments as the primary means of global AMR governance is questionable, and looks to how global governance deals with environmental issues for potential lessons.

The governance of non-health issues that have important impacts on health is the subject of the chapters by Meri Koivusalo and Caroline Thomas. Koivusalo provides a succinct introduction to how the specific provisions under various multilateral trade agreements of the World Trade Organization have policy implications for health. Through a broad ranging review of often complex and highly technical matters, she shows convincingly that governance for global health cannot be limited to strictly health-related institutions. Similarly, Thomas examines the potentially adverse effects of the Agreement on Trade Related Intellectual Property Rights (TRIPS) on access to essential drugs, notably in low-income countries. As demonstrated in the legal case in South Africa, the economic rationale driving large pharmaceutical companies to expand their markets globally can directly conflict with principles of health equity. Equitable access to essential drugs is critical to the protection and promotion of global health and the issue demonstrates the need to locate health more centrally in future governance of the trading system.

The book is usefully concluded by Ilona Kickbusch who neatly summarizes some of the key theoretical considerations for developing what she calls 'the new political space' of global health governance. Her chapter captures the sense that we are at a definitive turning point in health development and co-operation. There is rapidly accumulating evidence that 'something' is happening to patterns of health and disease and to the determinants of health. This 'something' is eroding and redefining familiar boundaries that have previously separated us spatially, temporally and cognitively. We urgently need to reflect on traditional ways of protecting and promoting population health centred on the state and the international states system.

There are encouraging signs that this is beginning to happen, but it remains unclear whether we are entering a 'power vacuum or a new political space'. We hope that this book will stimulate critical debate of existing and proposed forms of global governance in relation to their health impacts.

Note

1. It is estimated that 40,000 articles in the field of medicine are published each month.

Part I

The Impacts of Globalization on Health

2
A Global Political Economy Approach to AIDS: Ideology, Interests and Implications

Kelley Lee and Anthony Zwi

2.1 Introduction

The AIDS (Acquired Immune Deficiency Syndrome) pandemic has been, in many respects, unprecedented. The rapid emergence of a new disease and the mobilization of significant financial and human resources for research, treatment and control strategies, has seen no parallel.[1] Given the global spread of infection, and the global nature of the response, AIDS is of clear relevance to scholars of international relations (IR) and international political economy (IPE). Yet little attention has been devoted to health in the IR field, and even less to AIDS.

The bulk of initial research on AIDS in the early 1980s was devoted to understanding its occurrence and presentation within and across populations. In the past twenty years, there has been a growing body of research in the medical and social sciences focused at the national or subnational level. These studies have contributed much to exploring transmission of the human immunodeficiency virus (HIV) among particular groups (notably the homosexual community and intravenous drug users in the US and Europe, and commercial sex workers in sub-Saharan Africa, Asia and Latin America), the historical and cultural factors which have facilitated transmission of the HIV virus, and the economic impact of AIDS on health systems and labour markets. Other studies have examined individuals in terms of their sexual behaviour and awareness of HIV and AIDS. Analyses in recent years have broadened to include women and youth, gender, human rights, the impact of prevention strategies and the costs of treating HIV/AIDS.

As part of this broadening perspective on AIDS, this chapter argues that a global political economy (GPE) approach offers further insights for understanding and responding to the disease.[2] First, widening the spatial dimension of AIDS research to the global level is apt given the spread of the disease across national boundaries and the attempts to tackle it as a global phenomenon. Second, a GPE approach enables us to extend our analysis temporally by locating AIDS within a particular world order of 'transnational neoliberalism' defined below by material, institutional and ideological forces distinct to international relations since the late 1960s. After describing the global spread of AIDS, this chapter argues that there is growing evidence that this world order has created fertile conditions within which AIDS has spread, particularly among peripheral interests in the GPE. Importantly, this includes 'the way in which ideas about what constitutes the *political* and the *economic* have emerged historically' (Gill and Law, 1988, p. xviii – emphasis in the original), and it will be shown that policymaking on AIDS has been strongly centred on biomedical and neoliberal economic discourses.

The chapter concludes by arguing for a more critical awareness of the theoretical and practical lineage of existing knowledge of AIDS. Given that the limitations of this knowledge have been part of the conditions within which the disease has thrived, an expanded research agenda is put forth for the study of IPE and AIDS. It is in locating the disease within what Fortin calls a 'longer view', which seeks 'to situate the present disease within the broad historical, social, ecological, cultural, and political relationships that mediate humanity, sickness, and the environment' (Fortin, 1990, p. 217), that the beginnings of a truly global response to AIDS can be built.

2.2   The global spread of AIDS

AIDS is a complex of symptoms and signs in infected individuals, ultimately culminating in the fatal depression of the immune response system. The disease was first identified in 1981 among male homosexuals in the US. Within a remarkably short period the responsible infectious agent, HIV, was identified and the routes of transmission clarified. These were unprotected penetrative sexual intercourse, transfusion with infected blood products, injection using infected equipment and from an infected mother to her child during pregnancy, the birthing process or breastfeeding. There are two recognized serotypes of the virus, HIV-1 and HIV-2. HIV-1 predominates worldwide, while HIV-2 has spread since the mid-1980s, notably in west Africa, but by no means confined to that area.

HIV infection has spread rapidly across the globe. In North America, western Europe, Australia, New Zealand and many urban areas of Latin America, it is believed that HIV began to spread from the mid to late 1970s and early 1980s. This spread occurred initially among homosexual and bisexual men through unprotected sexual intercourse, among intravenous drug users (IVDU) and among the recipients of contaminated blood products. In sub-Saharan Africa, Latin America and the Caribbean, the heterosexual spread of HIV was recognized as a key means of transmission by the second half of the 1980s. Subsequent spread through infected blood products, given a high prevalence among the general population and limited access to technologies for screening and testing, is also likely. The estimated prevalence of HIV infection in these regions is sobering. In Africa, around 25 million adults are believed to have been infected to date, with over three million cases of AIDS.

Those at particular risk, such as commercial sex workers without access for instance to condoms, have been found to have HIV prevalence rates as high as 90 per cent (Hawkes and McAdam, 1993). In eastern and central Europe, the Middle East, north Africa, Asia and the Pacific, future increases in HIV infection are likely to be most dramatic, with the number of new cases of HIV infection in Asia to surpass those in Africa. Initial cases in Asia have occurred primarily among commercial sex workers and drug injectors. However, this pattern is rapidly changing and alarming increases in transmission and disease have occurred among the general population of both Thailand and India (Ford and Koetsawang, 1991; *The Lancet*, 1994). This looming epidemic is expected to 'ultimately dwarf all others in scope and impact' (Merson, 1993).

Table 2.1 shows that there was an estimated 34 million adults infected with HIV worldwide by the end of 1999. In some parts of North America,

Table 2.1 HIV/AIDS estimates and data, end 1999

Region	People living with HIV/AIDS, end 1999				
	Adults and children	*Adults (15–49)*	*Adult rate (%)*	*Women (15–49)*	*Children (0–14)*
Global total	34 300 000	33 000 000	1.07	15 700 000	1 300 000
sub-Saharan Africa	24 500 000	23 400 000	8.57	12 900 000	1 000 000
East Asia and Pacific	530 000	530 000	0.06	66 000	5 200
Australia and New Zealand	15 000	15 000	0.13	1 100	190
South and South-East Asia	5 600 000	5 400 000	0.54	1 900 000	200 000
Eastern Europe and Central Asia	420 000	410 000	0.21	110 000	15 000
Western Europe	520 000	520 000	0.23	130 000	4 100
North Africa and Middle East	220 000	210 000	0.12	42 000	2 000
North America	900 000	890 000	0.58	180 000	11 000
Caribbean	360 000	350 000	2.11	130 000	9 600
Latin America	1 300 000	1 200 000	0.49	300 000	28 000

Source: UNAIDS, Report on the global HIV/AIDS epidemic – June 2000.

Table 2.2   High, medium and low variant projections of the cumulative number of HIV infections, AIDS cases and AIDS deaths (millions) among adults in 2005, by region

	sub-Saharan Africa	Asia	North America	Latin America	Western Europe	World total[a]
Cumulative HIV infections						
High variant	23.3	23.0	1.6	5.2	1.0	54.7
Medium variant	20.7	19.1	1.5	4.6	0.9	47.4
Low variant	18.1	15.3	1.4	4.0	0.9	40.1
Cumulative AIDS cases						
High variant	9.4	5.3	1.0	2.0	0.6	18.5
Medium variant	9.1	4.9	1.0	1.9	0.6	17.8
Low variant	8.9	4.6	1.0	1.9	0.6	17.2
Cumulative AIDS deaths						
High variant	8.9	4.8	1.0	1.9	0.5	17.2
Medium variant	8.7	4.5	1.0	1.8	0.5	16.8
Low variant	8.5	4.2	1.0	1.8	0.5	16.2

Note:
[a] Total includes a small number of infections in the former Soviet Union, Eastern Europe, west Asia, and North Africa.

Source: John Bongaarts, 'Global Trends in AIDS Mortality', *Population and Development Review* 22(1): March 1996, p. 32.

western Europe, Australia and high prevalence areas of East and Central Africa, there is emerging evidence that the spread is stabilizing.[3] The number of heterosexual people infected with HIV in industrialized countries is expected to remain stable or increase more slowly as those currently infected die and new cases are averted (Cadwell, 1995). In other regions, however, the rate of increase will remain alarming high. Table 2.2 shows the projections of the cumulative number of HIV infections, AIDS cases and AIDS deaths in millions among adults in 2005, by region.

2.3   AIDS and the global political economy: an emerging transnational neoliberal order

There are no simple answers to how and why HIV infection has spread in the manner it has. Billions of US dollars have been spent on understanding the HIV virus, the pathogenesis of HIV disease, its epidemiology and clinical manifestations. Between 1982 and 1991, about US$5.63 billion was spent on AIDS-related research worldwide, the vast majority by the governments of the ten largest industrialized countries and almost all since 1985.[4] Total global research and development for preventative HIV vaccines in both the public and private sectors in 1999 was in the order of only US$300 million (IAVI, no date).

In order to broaden our understanding of the AIDS pandemic, this chapter examines how the study of the global political economy can enhance our understanding of HIV and AIDS. In turn, it is argued that HIV and AIDS can reveal new empirical applications for IPE scholars. Traditionally, the realist–liberal orthodoxy, which has dominated the study of international relations, has neglected the study of health. As Thomas writes, 'disease is a transnational phenomenon which pays no heed to territorial state boundaries; yet it rarely features in the discussion of International Relations' (Thomas, 1989). Critical or post-positivist approaches, which have emerged, seek to widen the intellectual boundaries of IR, extending the disciplinary agenda to include gender, the environment and social policy (Murphy and Tooze, 1991).

Briefly, the GPE approach taken in this study is based on the writings of critical theorists in IR. Critical theorists share the aim of challenging positivist approaches to social science and proposing alternatives. Applied to the GPE, Gill writes that, first, 'there is a relativity in the claim to truth'; and second, that 'social conditions interact with and influence the survival, scientific status and consequences of rival social theories: knowledge is also a process of social struggle, again between hegemonic and counter-hegemonic perspectives and principles'. Given that all 'thought processes and knowledge systems' are at variance from reality, the aim is not to determine which truth is a true reflection of reality, but 'how and why and with what consequences' these particular 'truths' hold within a given historical period (Gill, 1993, p. 45). In this sense, the approach goes beyond the economic determinism found in early political economy of health literature (Doyal, 1979; Navarro, 1981).

This dialectic between the realms of ideas and material conditions is central to the ontology and theory of GPE put forth in critical theory. Drawing on the writings of Antonio Gramsci, Gill writes that the global political economy 'implies an integrated system of knowledge, production and exchange, and includes the dialectical relations between capital and non-capitalist systems and states, and ecological, ethical and other aspects of the whole' (Gill, 1993, p. 45). Central to such relations is the Gramscian concept of *hegemony* that is achieved by core interests (individuals, classes, states or transnational actors) through a mixture of coercive and consensual means. The use of coercion or force to assert leadership is domination. In contrast, the mobilization of the consent of peripheral groups to a given order is achieved through intellectual, moral and political leadership. This suggests a complex concept of power that can be structurally embedded in historical social constructs (that is historic blocs), as well as vested in individual actors (Lee, 1995).

Applying this GPE approach to the study of AIDS, it is argued that the disease can be located within a particular world order characterized by transnational neoliberalism. Briefly, this world order is defined by unprecedented growth of capital, production and exchange relations across state boundaries, with the effects of this process of globalization[5] far from uniform. The material basis of this world order is 'a global economic system dominated by large institutional investors and transnational firms which control the bulk of the world's productive assets and are the principal influences in world trade and financial markets.' Its institutional nucleus is comprised of 'elements of the G-7 state apparatuses and transnational capital (in manufacturing, finance, and services)', which have created an 'internationalization of authority and governance that not only involves international organizations…and transnational firms, but also private consultancies and private bondrating agencies' (Gill, 1995). Global governance is thus led by what Cox calls a 'transnational managerial class' and their institutional homes (for example ministries of finance, trade and industry, the World Bank and IMF, the Organization of Economic Cooperation and Development), who increasingly shape policymaking on a global scale (Cox, 1987, p. 359). Intellectual, moral and political leadership of this transnational managerial class is centred on consent to the ideas of political and economic neoliberalism, an ideology that is 'largely consistent with the world view and political priorities of large-scale, internationally-mobile forms of capital'.

It is within this world order that a fuller understanding of the global spread of AIDS can be achieved. First, the spread of AIDS has been facilitated by resultant changes in the spatial dimension of human relations. Globalization, as described above, has led to an unprecedented increase in the worldwide movement of people (within and across national borders) in the form of migration and migrant labour (Bassett and Mhloyi, 1991), tourism, displacement, occupying military forces[6] and rapid urbanization. Where development strategies have been adopted which seek to integrate a country into the global economy, through liberalized markets and rapid industrialization, pressures of transience on populations have increased at the same time as weakening systems of social support and the increased prices of many products and services. These changes have led to fundamental changes to social structures in both lower and higher income countries that have provided fertile conditions within which HIV has appeared and spread. For example, in Senegal and Uganda, it was found that 'mobility is an independent risk factor for acquiring HIV', noting that 'it is not the origin, or the destination of migration, but the social disruption which characterizes certain types of migration, which determines vulnerability to HIV' (Kane *et al.*, 1993; Decosas *et al.*, 1995). Similarly, a study in the US found that 'the disintegration of minority community physical or social structures has created such turmoil that AIDS spread rapidly to intravenous drug users and their sex partners, and rapidly followed commuting patterns into the suburbs' (Wallace and Wallace, 1994).

Second, the ability of individuals, societies and countries to adapt to the process of globalization is unequal, with those less able to adapt engaging in risk behaviours under conditions of poor access to health care, that have made them more susceptible to HIV infection. For many low-income countries, the period since the late 1970s has brought rising levels of foreign debt, negative economic growth, declining terms of trade and levels of foreign aid. The introduction of structural adjustment programmes (SAPs) by the World Bank during the 1980s as a response to these macroeconomic problems has, as Asthana writes, 'contributed to an unprecedented decline in the health and living standards of the Third World poor' (Asthana, 1994, p. 51). The main cause has been reduced public spending on the social sectors, which much recent evidence shows has led to a decline in health status among the poorest communities. In Zambia, real expenditure on health fell between 1982–85 by 22 per cent, while in Bolivia per capita expenditure on health by the government declined by 30 per cent (Asthana, 1994, p. 54; Stewart, 1992). The introduction in many countries of cost recovery mechanisms, such as user charges for health services, has been found to reduce utilization despite exemption mechanisms for the poor (Gilson *et al.*, 1995). It is now widely recognized by the development community, including the World Bank, that in the mid 1990s, insufficient emphasis on 'poverty alleviation' by SAPs has had detrimental effects on the health status of poor communities (Ugalde and Jackson, 1995).

It has been within communities worst affected by these declining health conditions that HIV infection has gained the biggest foothold. The fastest growing epidemics have been among the most socioeconomically disadvantaged populations within and across countries. In low-income countries, greater human insecurity and social inequities have led many individuals to resort to coping strategies (for example migrant labour, commercial sex) under conditions that have placed them at greater risk of HIV infection. In Tanzania, it was found that 70 per cent of commercial sex workers engage in prostitution for economic survival (Panos Institute, 1988, p. 88). In west Africa, large numbers of men migrate seasonally to work in neighbouring countries because of poor employment opportunities at home. In a study of Côte d'Ivoire, it was found that an agricultural enterprise provides accommodation for 2000 young male migrant workers in a camp that is visited after each pay-day by sex workers. Each sex worker 'services' a 'mean of 25 workers each over a period of two nights'. The study found that it is 'this pattern of mixing, and not the actual number of partners, which is responsible for the extensive spread of HIV among migrant workers' (Decosas *et al.*, 1995, p. 827). In India, research by the WHO found that poverty is a key factor contributing to AIDS because financial hardship has forced men to leave their families to find employment. The use of drugs and/or sex workers has often followed suit (Lal, 1995). An exacerbating factor in the spread of HIV infection among these low-income groups is the inadequate treatment of sexually transmitted diseases (STDs). Conditions such as genital ulceration increase the vulnerability to HIV infection, yet treatment for such conditions has not been forthcoming given limited access to health services. A study in Nairobi showed declines in clinic attendance for sexually transmitted diseases following the introduction of user charges (Moses *et al.*, 1992).

It has not only been in low-income countries that deteriorating political economies have increased vulnerability of populations to HIV infection. In the 'economies in transition' of central and eastern Europe, the end of the Cold War and subsequent pressures to integrate rapidly with the emerging transnational neoliberal order, have brought upheaval to political and economic infrastructures. Health systems have been particularly badly affected, accompanied by sudden rises in such health indicators as infant mortality rate, incidence of infectious diseases and substance abuse. In relation to AIDS, the incidence of high-risk behaviours within unsafe conditions has increased (Danziger, 1996; Field, 1994). In Albania, described as the 'bargain-basement brothel of the West', 'sex tourism' has become one of the main sources of income for many women (Einhorn, 1993, p. 136).

Finally, in many western industrialized countries, globalization has led to the rise of a disenfranchized, disempowered and impoverished 'underclass'. Individuals and groups within this class are those most adversely affected by globalized economic and political relations: the under- or unemployed thanks to monetarist economic policies; low-skilled labour because of the international mobility of production; small businesses because of the globalized scale of production of goods and services; and women and ethnic minorities who make up the majority of part-time and temporary workers. In the UK this class has emerged amid the widest social inequalities since the 1930s (Hutton, 1995; Timmons, 1995). Similarly, Singer (1994) found that urban poverty and socially devalued ethnicity have been factors contributing to the spread of HIV/AIDS in the US.[7]

In summary, there is much evidence that the emergence of a transnational neoliberal order has widened the health gap between haves and have-nots within and across countries. Within both lower and higher income countries, individuals and groups have had variable capacity to respond to rapidly changing global political and economic relations. Across countries, Figure 2.1 shows that a small number continue to control the vast majority of global wealth. Over 80 per cent of people estimated to be infected with HIV live in low-income countries (World Bank, 1993, p. 99). While some have gained access to employment, education and quality health care, others have faced limited employment and educational prospects, social dislocation and limited access to health care. It is within conditions of poverty

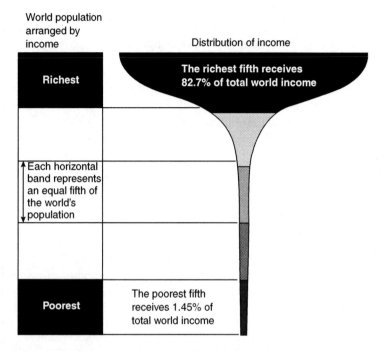

Figure 2.1 Champagne glass of inequality in global wealth

Source: N. Alexander, 'Remarks to the Banking Subcommittee on Domestic and International Monetary Affairs, US House of Representatives on the World Bank and Poverty', Bread for the World Institute, 27 March 1995.

that low-income groups have been more likely to engage in risk behaviours that heighten their exposure to HIV infection.

2.4 Orthodoxy and ideology: the discourse of HIV/AIDS

It is within the context of the global inequities described above that a critical analysis of the discourse on AIDS can be located. As well as changes in the structural conditions in which AIDS has emerged and spread, a key aspect of the GPE of AIDS is the way in which the disease has been understood by both scholars and practitioners. How AIDS has been theorized has had profound implications, not only for how we understand the disease, but for our responses to it. At the global level two main discourses, biomedical and neoliberal, have dominated (for a discussion of how the biomedical and economistic discourses have shaped policymaking at the national level see Seidel, 1993).

The *biomedical discourse* has focused on the clinical and epidemiological characteristics of the disease, modes of transmission, strategies emphasizing individual risk behaviours and developing methods of prevention, treatment and care. The proponents of this biomedical discourse have been scientists and medical professionals, trained within the Western medical tradition, who see the origins of disease primarily in terms of individual agents (that is viruses, people) and seek solutions through the medical sciences directed at the agents of infection. This tradition, as Doyal writes, is 'hospital-centred, highly technological, and is dispensed on an individual curative basis' (Doyal, 1979, p. 255; see also Illich, 1976; Moon and Gillespie, 1995). For AIDS, the problem is seen to be a lack of knowledge and its application. It is held that, with sufficient amounts of the 'right' information, AIDS can be 'solved'. So far, the search for this 'magic bullet' to treat and cure AIDS has been unsuccessful (Doyal *et al.*, 1994). Policymakers in the meantime have turned to promoting preventive measures, such as education and information campaigns, directed at changing individual behaviours.

Critiques of the biomedical approach to health are not new. Of particular note has been Lesley Doyal's *The Political Economy of Health* (1979) in which she argues that the legacies of early capital accumulation, social inequalities, colonialism and racism laid the foundations for ill-health in both developed and developing countries. Other writers, such as Vincent Navarro (1981), David Sanders (1985) and Meredith Turshen (1989) have raised similarly important questions for social scientists seeking to understand the relationship between medicine, health and society. These ideas have increasingly been applied to HIV/AIDS, albeit with emphasis on the national and subnational levels (Cadwell, 1993; Evian, 1994; Sanders and Sambo, 1991; Zwi and Cabral, 1991).

In relation to AIDS, there are three main shortfalls to the biomedical discourse. First, by focusing on the biology of the individual, it neglects potential explanations of disease and its treatment within wider social, economic and political contexts that create conditions of risk. Second, the discourse conceives the major response to health problems in terms of medical care and thus fails to recognize the key importance of power in shaping policy responses to AIDS. Third, because medicine is seen as essentially 'good', the problem is perceived in terms of insufficient amounts of medical knowledge rather than any shortfalls in the knowledge itself.

A second and increasingly powerful AIDS discourse since the mid-1980s has been based on the principles of *neoliberal economics*. The ascendance of neoliberalism within the health sector has been part of the wider global shift in ideology, interests and institutions described above. Neoliberalism argues that the causes of ill-health are allocative and technical inefficiencies, and that health can be improved through, for example, reducing corruption and waste, supporting cost-effective interventions and increasing the operation of market forces in health service delivery and financing.[8] Like the biomedical discourse, it implicitly assumes that ill-health is primarily a resource issue, in this case, making better use of health resources.

Applied to AIDS, this discourse has focused on the financial implications of the disease for national economies, with direct costs estimated at US$11 billion worldwide in 1993 (Mann and Tarantola, 1995). Once AIDS infection rates surpass 5 per cent, the impact on gross domestic product is measurable and prevalence rates of more than 20 per cent can knock as much as 2 per cent off annual GDP growth. For example, in Botswana, HIV/AIDS will reduce potential GDP by nearly one-third by 2025 (Williams, 2000). Costing studies have thus been carried out, for example, on the impact of AIDS on patterns of health care expenditure and the cost-effectiveness of different control strategies (Ainsworth and Over, 1992; Cuddington, 1993; Foster and Buve, 1995). Other studies have examined the effect of the disease on local labour markets, given that, in many countries, AIDS has affected people in their most economically productive years (World Bank, 1993, p. 100).

The main shortfall of the neoliberal discourse has been the legitimation of economic above other criteria (for example moral, ethical, clinical) in the allocation of resources for AIDS. Within the context of seeking greater 'value for money', an increasing number of national and international policymakers have placed financial considerations foremost in the selection of control and treatment strategies. The promotion of equity, human rights and gender issues have received lower priority. This is perhaps best illustrated by the World Bank's *World Development Report 1993: Investing in Health*, which puts forth the concept of the 'global burden of disease' measured by 'disability adjusted life years' (DALYs). A DALY is calculated by 'the present value of the future years of disability free life that are lost as a result of the premature deaths or cases of disability occurring in a particular year'. Importantly, this formula is based on the normative assumption that individuals in economically productive age and socio-economic groups are of higher value. AIDS is calculated to cause a heavy burden of disease and, hence, should be given high priority by policymakers, because it has 'powerful negative economic effects on households, productive enterprises, and countries' (p. 100).

Together, both the biomedical and neoliberal discourses have framed the AIDS debate in apolitical terms, legitimized by the claim that they are confined to non-normative issues. Little discussion has taken place within these discourses of the role of power and particularly how AIDS can be located within changing structures of global political and economic power. Furthermore, the dominance of these discourses has not been limited to an intellectual exercise, but has directly shaped the global response to AIDS.

2.5 The global response to AIDS: policymaking and the biomedical discourse

National and international policymakers were initially slow to respond to AIDS as a public health issue. In high-income countries this was largely the result of its characterization by the medical community as being limited to, and controllable within, particular 'risk groups' (homosexuals, IVDUs, certain geographical regions). Between 1981 and 1985, researchers focused on gathering clinical, albeit vital, information on the disease. This slow response by public health officials was exacerbated by distrust within affected communities. Within the Afro-American community, which has suffered disproportionately from HIV infection in the past decade, efforts by public health officials were looked upon with suspicion because of historical experiences of unethical medical research practices. Most notorious was the Tuskegee study on syphilis in which researchers left people untreated to allow for observation of longer-term effects (Thomas and Quinn, 1991). In sub-Saharan Africa, apparent Western preoccupation with the alleged link between Africa and AIDS was seen by many as racist and delayed recognition and public acknowledgement of the importance of HIV infection in some regions (Chirimuuta and Chirimuuta, 1989).

It was not until 1985 that the first International AIDS Conference was held, after which there occurred a mobilization of expertise and other resources with unprecedented speed. Soon afterwards the international community began to discuss AIDS as a serious global issue and, in February 1987, WHO's Global Programme on AIDS (GPA) was created as the main multilateral channel for Western aid (Mann, 1988). This was followed by a high-level consultative meeting on AIDS and Development held in Talloires, France, at which attendance included the heads of WHO and the UN Children's Fund (UNICEF), senior aid officials and the Minister of Health for Uganda, a country where AIDS was already well-established. In 1987 AIDS became the first disease to be discussed at the UN General Assembly, which prompted Resolution 42/8 to mobilize the entire UN system in a worldwide campaign to control the disease. A World Summit of Health Ministers on AIDS was held soon after in January 1988, which put forth the London Declaration on AIDS.

Financial resources for AIDS have reflected this rapid growth in international attention. In 1987 US$30 million was allocated by GPA; by 1990 this had grown to US$109 million. Total external aid flows for AIDS and STD reached US$185 million in 1990, more than ten times that for tuberculosis and 20 times that for intentional injuries, conditions responsible for a similar 'burden of disease' (Michaud and Murray, 1994). By 1991 global spending on AIDS reached 10 per cent of total investments in health and population, with development agencies contributing some US$250 million annually (Decosas and Finlay, 1993), with notably, 95 per cent of annual global AIDS budgets spent in high-income countries (*The Lancet*, 1994). By 2001, US$1.8 billion was spent on AIDS globally (Boseley, 2001), and US$9.2 billion will be needed by 2005 annually to support an expanded response to HIV/AIDS in low- and middle-income countries (Schwartlander *et al.*, 2001).

Bilaterally, the US has spent the largest sums on HIV/AIDS. By the end of 1993, the US government had spent over US$17 billion, with nearly US$1 billion on basic scientific research by the Public Health Service in 1993 alone (USAID, undated). Annually, the US spends about $460 million on HIV/AIDS internationally, accounting for nearly half of international expenditure worldwide (Barks-Ruggles, 2001). Abroad, the US Agency for International Development (USAID) has made the highest financial contributions to HIV prevention activities in low-income countries. Initially officers were advised to minimize their involvement in HIV/AIDS, given limited available resources and sensitivity over the distribution of condoms. It was only after recognition of the impact of HIV/AIDS on other development efforts (for example child survival) that the US played a more active role in the creation of GPA, providing US$88.5 million of funding from 1986–91. This was in addition to US$79.5 million allocated by USAID for bilateral projects and another US$68 million worth of other USAID activities supporting HIV prevention (US General Accounting Office, 1992).

From a GPE perspective, this growth of international effort to control HIV/AIDS from the early to mid-1980s was firmly located within the biomedical discourse. Internationally, GPA was designated as the lead organization to mobilize resources and promote the development of national AIDS control programmes. Nationally, governments worked closely with GPA to initiate and implement their own strategies. These strategies were centred on the screening of blood products, clinical care for those affected, education and information campaigns, social marketing of condoms and peer education among commercial sex workers. The screening of blood products was particularly emphasized during the early years, despite being the least common route of infection (GTZ, 1995). As a technical intervention, it was seen as less controversial than, for instance, the promotion of condom use. Improving the safety of blood supplies also had other potential benefits, such as safety from other blood-borne infections. By no means insignificant were the concerns among travellers from high-income countries to the developing world (Buchanan, 1987; Potts and Carswell, 1993). Information and education campaigns warning of 'risk behaviours' have also been an important component of prevention strategies. In these campaigns, emphasis is placed on informed individuals taking responsibility for not engaging in such behaviours. HIV/AIDS was seen, in this context, as a disease of poor technical capacity, clinical practice or personal knowledge, and less a disease of broader structural features within and across countries.

As national HIV/AIDS control programmes became widely adopted by governments, and resources began to flow to support their implementation, many policymakers turned to the development of appropriate drugs to treat the disease and a vaccine to prevent HIV infection as key components of a global AIDS policy. With the prominent success of WHO's Intensified Smallpox Eradication Programme still in recent memory and other campaigns such as the Polio Eradication Initiative underway, many believed that the AIDS pandemic could also be defeated if the right technologies were available. Thus the race to develop such technologies has led to a massive effort by researchers worldwide. In a survey of 'leading [vaccine] researchers, public health officials, and manufacturers' carried out by the magazine *Science* in 1994, a vaccine for HIV infection was ranked first as the 'most urgently needed vaccine' in both high- and low-income countries (*Science*, 1994).

While biomedical research has and will continue to contribute vital knowledge in the treatment and control of AIDS, social scientists have played a prominent role in pointing to the need to take account of broader political, economic and social factors behind the AIDS pandemic. In the mid-1990s, there is much concern that biomedical interventions alone have failed to halt the alarming spread of the disease. Policymakers have responded by reassessing the global AIDS strategy. In January 1996 GPA was replaced by UNAIDS, a joint and co-sponsored programme of the WHO, UNICEF, UN Population Fund (UNFPA), UN Development Programme (UNDP), UN Educational, Scientific and Cultural Organization (UNESCO) and the World Bank. Foremost among the new programme's goals has been to develop a multisectoral and multidisciplinary approach to HIV/AIDS, drawing from a broader range of knowledge and expertise. How well this has been achieved, amid uncertainty and rivalry among UN organizations, is unclear. Yet it reflects a recognition that the biomedical discourse alone cannot provide a complete global policy response to AIDS.

2.6   The global response to AIDS: policymaking and the neoliberal discourse

As well as the biomedical discourse, the global response to AIDS has been strongly shaped by neoliberalism. As described above, neoliberal economic principles such as market-driven resource allocation, privatization, deregulation and free trade have influenced national and international health sector reform since the 1980s. In relation to AIDS, it is argued that the neoliberal discourse has legitimized certain institutions, interests and ideas in the setting and evaluation of policies according to economic criteria. First, the rise of neoliberalism within the health sector during the 1980s was accompanied by the increased prominence of the World Bank in international health policy. Since 1980, when the Bank began to lend specifically for health projects, it has become the largest source of external health financing for low-income countries. This financial clout has been followed by a greater voice in policy development, with Bank publications having widespread influence on the international health policy agenda (Buse, 1994).

World Bank lending for AIDS projects began in 1986 and by 1993 it had loaned a cumulative total of US$500 million (World Bank, 1994). Between 1996 and 2000, the Bank committed approximately US$493 million for new HIV/AIDS components and stand-alone projects in 39 countries.[9] One of the key activities supported by the Bank has been research on the economic effects of HIV and the cost-effectiveness of alternative interventions. Lending policies for AIDS programmes, in turn, have been based on the argument that prevention among 'high-risk groups' is one of the most cost-effective health interventions in the 'global burden of disease' (World Bank, 1993). For example, in India the Bank has worked with the government to study the cost of health care and the indirect national income loss from AIDS. As part of a US$84 million 'multi-pronged prevention strategy', the Bank has combined this research with biomedical interventions emphasizing 'behavioural change and condom use, control [of] STDs, strengthen[ing] blood safety, rais[ing] surveillance and clinical management capacity, enhanc[ing] national and provincial capacity for managing HIV/AIDS control activities, and ensur[ing] humane treatment of people with AIDS or HIV infection' (World Bank, 1994). Similar studies have been carried out in the worst affected countries in sub-Saharan Africa and in Thailand. Overall, the World Bank's lending policy on AIDS has been strongly informed by the neoliberal discourse. Neoliberal economic principles have been foremost in the allocation of resources for AIDS over other health needs and for the selection of specific interventions. These criteria have been backed by large amounts of financing by the Bank.

Second, the neoliberal discourse has led to an increased emphasis on non-state health care financing and service delivery as viable alternatives to nonexistent or weakened government institutions. As part of a broad shift in the 'public–private mix' within the health sector of many countries, there has been a marked increase in the role of both 'for-profit' (for example transnational corporations) and 'not-for-profit' institutions (for example NGOs) in health since the mid-1970s. In low-income countries, this trend has been partly fuelled by an increase in health sector aid through NGOs. Between 1975 and 1985, total official development assistance (ODA) grew by 115 per cent with the amount channelled through NGOs increasing by 1400 per cent. By 1990 NGOs accounted for 17 per cent of the US$4800 million disbursed as external assistance (Michaud and Murray, 1994), and about 13 per cent of health sector aid (Lob-Levyt, 1990). The rationale for this support includes the belief that NGOs can enhance political and economic pluralism (civil society), work more closely with local communities and marginalized groups and fill gaps in service provision. NGOs are also seen to encourage greater cost-effectiveness by offering alternatives to inefficient public health care, and even creating a competitive market where greater choice and competition will improve services (Bratton, 1989; Crane and Carswell, 1992). There is no doubt that NGOs are playing a vital and, in many cases, groundbreaking role in the field of AIDS. They have, for example, been at the forefront in drawing attention to human rights issues. The effective implementation of community-based strategies will also depend in large part on NGOs. However, questions have also been raised about the lack of accountability of these organizations and their financial links with traditional donor agencies. There are also concerns of duplication or the establishment of parallel structures, some of which undermine the central role of government (Godwin, 1995; Green and Matthias, 1995).

Third, the neoliberal discourse has strongly defined the race for biomedical technologies to treat and prevent AIDS. The key players in this race have been large pharmaceutical and biotechnology companies, such as the US companies Repligen Corporation, Merck & Company, and MicroGeneSys Incorporated, which have spent billions of dollars since the 1980s on research and development. While public subsidies helped to motivate companies into establishing research programmes,[10] the main spur has been the prospect of a profitable payoff should an effective treatment or vaccine be found. A similarly lucrative market for blood-testing equipment and storage has already emerged.

Leaving the development and presumably the eventual allocation of potential treatments and vaccines to the private sector raises several concerns, including the question of access. Private companies have clearly identified their key markets to be in high-income countries where individuals and governments can afford to pay the expected price. An estimate put forth in 1990 of the potential market for vaccines in the US and Europe was 67 million people. It was calculated that, if 15–20 per cent of these people paid US$150, this would create a market of over US$1.6 billion (Cohen, 1994). Today, the global market for all vaccines is estimated to be US$3.8 billion (*IAVI Report*, 2000), although the potential market for an AIDS vaccine is perceived to be small and confined to OECD countries (IAVI, undated). From the companies' perspective, it would require such a large return to pay for the huge sums that pharmaceutical companies invest in the research and development of new products. As argued by the Vice President for Scientific Affairs of the International Federation of Pharmaceutical Manufacturers Associations (IFPMA), Margaret Cone, 'if companies are going to invest enormous sums to develop a vaccine, they have got to get a good return on their investment. And an AIDS vaccine is not going to be cheap. The manufacturing procedure is likely to be long, complicated and expensive' (quoted in WHO, 1992a). In low-income countries, however, where more than 80 per cent of people infected with HIV currently live, and where 90 per cent of new infections currently occur, this 'market price' lies beyond the reach of the vast majority of people. For example, Uganda is among the countries worst affected so far by HIV and AIDS. An estimated 1–1.5 million people are HIV infected and GNP per capita is US$170 (World Bank, 1993). For many Africans, a vaccine costing US$150 will clearly be economically inaccessible.

A related concern is the danger of over-reliance on market-driven research should the perceived market prove unprofitable. Indeed, in the mid-1990s, some private companies began to reassess the investment potential of AIDS vaccine research in the light of new estimates of market potential in high-income countries (McCarthy, 1995). Recent evidence has shown that rates of HIV infection have stabilized in some high-income countries, while they continue to rise in many low-income countries. In July 1994 the US bio-technology company Repligen Corporation announced the cancellation of its AIDS vaccine programme for financial reasons. Many are concerned that public health priorities, rather than profits, need to be behind decision-making over research.

A further concern lies in the clinical appropriateness of the research being carried out in meeting the needs of countries worst affected by HIV and AIDS. In the report of a meeting on HIV vaccines held by the Rockefeller Foundation at Bellagio, Italy in 1994, concerns were expressed that companies are 'catering to the needs of the developed world' by focus-ing on 'a small number of the potential vaccine approaches' (Rockefeller Foundation, 1994). This has included an emphasis on the HIV strain sub-type B, which predominates in Europe and the US but not in low-income countries. One notable example has been trials in China and Thailand beginning in June 1993 of a potential HIV-B vaccine. In China the trials have been carried out by the US company United Biomedical with agree-ment of the Chinese government, while in Thailand it has been a joint project between the US and Thai armies. Grady writes that governments and private companies have been under pressure to 'do something ... do anything' in the development of an AIDS vaccine (Grady, 1995). However, both projects have been criticized on the grounds that the HIV-B virus generally does not occur in Asia.

In summary, the global policy response to AIDS from the mid to late 1980s has been strongly informed by a neoliberal discourse. As well as con-ceptualizing the problem of AIDS in economic terms, neoliberalism has privileged the needs of certain interest groups, including private compa-nies, and has defined the criteria by which policy responses have been evaluated and implemented.

2.7   Towards an expanded understanding of AIDS: a global political economy research agenda

The purpose of this chapter has been to bring the study of the GPE and AIDS closer together and to show how each can be more fully understood through the other. We argue that, while social sciences have made vital contributions to the study of AIDS over the past decade, scholars of IR have largely neglected global health issues. This is beginning to be addressed (Gordenker *et al.*, 1995), but there are several avenues of research to be explored that would enrich our understanding of both the emerging GPE and the AIDS pandemic.[11]

First, there is a need to explore more fully the relationship between the emerging structure of global economic and political power, and the epidemiology of AIDS. As Porter writes, 'the profile of the epidemic is increasingly one of poorer people in developing countries... [yet] resource allocations are declining in many quarters' (Porter, 1995). The link between structures of national political economy and health has been increasingly explored, for example, in relation to the debt crisis, urban decay and perceptions of personal responsibility for health (Alubo, 1990; Donahue and McGuire, 1995). These analyses must now be extended to the global level by pulling together existing studies, as well as studying AIDS as a transnational and global phenomenon. What has been the relationship between the spread of AIDS and levels of foreign debt in low-income countries? Has there been a link between the employment practices of transnational corporations and the global pattern of AIDS? What impact has the movement of armed forces had on the spread of HIV infection? What aspects of globalization must international organizations, like WHO and the World Bank, recognize as contributing factors to the spread of AIDS? Are existing AIDS policies focused at the national level, sufficiently taking account of global factors? What can AIDS tell us about the winners and losers of globalization? Have changes in international relations since the end of the Cold War, such as a potential growth in global civil society and the rise of ethnic and territorial tensions, influenced policymaking on AIDS? How has the transformation of production relations, including the reorganization of labour and technological change, affected the health status of workers worldwide?

Second, more detailed and critical analysis of the epistemological and ontological lineage of policy responses to AIDS so far is clearly needed. This chapter has begun to show how the biomedical and neoliberal discourses have strongly informed the AIDS debate within key institutions since the early 1980s. Further study of the GPE of AIDS would show more precisely how these discourses have been located within, and disseminated by, the many other individuals and institutions that have become involved in AIDS research, treatment and control, education and policymaking. This includes the impact of global telecommunications and the mass media on how we think about and respond to AIDS. Identifying emerging contradictions between such thought and actions intended to prevent and control the disease forms a key part of this critical analysis. Furthermore, there is a need to consider what other discourses, from the rich history of political and social thought, can be explored to broaden the AIDS debate methodologically, theoretically and conceptually. Accompanying the existing orthodoxy has been limited recognition of the structural and historical roots of the pandemic, and hence the more fundamental social and political changes needed to address it effectively. AIDS, in short, teaches us about the current limitations in thinking about health, development and international relations.

Third, there is an urgent need to apply an expanded understanding of the GPE to develop ethical and practical approaches to AIDS. National and international policymakers are presently seeking to widen the AIDS agenda in recognition that biomedical and economistic responses alone are insufficient. As Mann and Tarantola write, 'it is now clear that HIV/AIDS is as much about society as it is about a virus. This new understanding of the societal basis for vulnerability to HIV/AIDS has the potential to provide a strategic coherence to efforts in HIV/AIDS prevention and control' (Mann and Tarantola, 1995). Public health experts are looking to social scientists to contribute to a new policy agenda (Walker, 1994), with hopes that there will be a move in focus 'away from science to issues of sociology and sexual behaviour, organizational and management of programs, NGOs and Government relations and the epidemic's economics' (Porter, 1995, p. 5). A key part of this new agenda must be the impact of globalization on health. It is on this subject that scholars of the global political economy can offer much. As part of our efforts to redefine the concept of security in a post-Cold War world, issues concerning global human security (that is health, environment, migration) need to be given far greater and urgent attention.

Acknowledgements

An earlier version of this chapter was presented to the Health and International Relations Panel, British International Studies Association Annual Conference, University of York, December 1994. Thanks to Charlotte Watts who provided useful comments on the penultimate draft of this paper, and Preeti Patel who updated the data presented since the paper's publication in *New Political Economy*, vol. 1, no. 3, 1996.

Notes

1. The other notable global effort to control or eradicate a specific disease is the WHO's Intensified Smallpox Eradication Programme (1967–80). The total cost of this programme was US$81 million. In comparison, the budget for WHO's Global Programme on AIDS was about US$100 million in 1989 alone.
2. This article uses the term '*global*' rather than '*international*' political economy because, as Gill and Law argue, 'political economy analysis should not be narrowly limited to relations between nation-states and their governments. It needs to be global as well as international in character' (Gill and Law, 1988, p. xii).
3. Stabilization is defined as the number of deaths from AIDS over the past year being roughly equal to the number of new cases of HIV infection. However, it is important to note that this measure may hide disproportionate increases in certain groups within a national population (WHO, 1994, p. 4).
4. Figures of amounts spent by pharmaceutical companies are not readily available (Mann *et al.*, 1992, p. 263).
5. It is recognized that the concept of globalization has been applied with 'casual abandon' by scholars and policymakers. It is defined here as a process of intensified

social, political and economic relations on a global scale. For a fuller conceptual discussion see Jones (1995).

6. The Civil–Military Alliance to Combat HIV and AIDS was established in November 1994 in recognition that the military tends to have sexually transmitted disease rates two to three times that of civilian age group counterparts. This rate multiplies even further in wartime (Groennings, 1995, p. 8).

7. Singer (1994) found that almost half of people in the US who have been diagnosed with AIDS are African-Americans and Latinos from impoverished urban neighbourhoods.

8. The seminal publication is the *Financing Health Services in Developing Countries: An Agenda for Reform* by the World Bank (1987).

9. By the end of 2001, the Bank will have committed an additional US$740 million for new HIV/AIDS prevention and care efforts, mostly in sub-Saharan Africa ('Turning the Tide Against HIV/AIDS Time for Global Action to Match a Global Threat', World Bank Press Release, 21 June 2001, www.worldbank.org/ungass/home.htm).

10. For example, the US government spent US$111 million out of US$160 million spent worldwide on HIV/AIDS vaccine research and development (see Cohen, 1994).

11. This proposed research agenda is partly based on the new historical materialistic research agenda for the study of global politics put forth in Gill (1995, pp. 16–17).

3
Understanding HIV/AIDS as a Global Security Issue[1]

Dennis Altman

> The defense this nation seeks involves a great deal more than
> building airplanes, ships, guns and bombs. We cannot be a strong
> nation unless we are a healthy nation.
>
> (US President Franklin Roosevelt 1940,
> as quoted in Fallows 1999, p. 68)

3.1 Introduction

In mid-1999 the South African government placed an order for three new
submarines at a cost of about US$680 million dollars (Hawthorne, 1999).
At the time, approximately 1500 people per day were becoming infected
with HIV. One might legitimately ask whether the money spent on building its naval forces would not more appropriately be spent on fighting
HIV/AIDS. As *The Economist* (1999) described, 'AIDS will probably kill
thousands of times more South Africans than apartheid ever did.' As such,
the country is a particularly appropriate example of the broader political aspects of the HIV/AIDS pandemic. On the one hand, the country has
experienced bitter domestic debates on the adequacy of the government's
response, centring on issues such as whether the government should supply AZT to HIV-positive pregnant women and, indeed, whether HIV is the
primary cause of AIDS (Marais, 2000; Schneider, 2001; van der Vliet, 2001).
On the other hand, the government has sought to engage with the broader
dimensions of the pandemic, including using the courts to battle attempts
by pharmaceutical companies to place restrictions on the availability of
generic drugs. Notably, this has not necessarily resulted in the sort of national
plan for access to both treatments and prevention associated with Brazil
(Rosenberg, 2001).

   If one of the primary aims of a state is to protect the lives of its citizens,
then risks to security can come in many forms other than those from conventional warfare. The discipline of international relations has gradually
been coming to terms with this argument, although with some reluctance,

given its continued dependence on seeing states and the relations among them as central. This, in turn, makes it difficult to see factors such as environmental degradation, refugee flows or infectious diseases as threats to security that might be comparable to bullets or supersonic missiles.[2] In contemporary attempts to redefine security, it is now common to list a number of issues (for example international terrorism, drug trafficking) to which issues of health generally, and epidemics of infectious diseases more specifically, are often added. Thus a survey of new international relations edited by Cooperrider and Dutton (1999, p. xv) pronounces in its opening paragraph:

> Today's global forces for change are moving us into a remarkable new set of circumstances, one in which human social organizations inherited from the modern era may be unequal to the challenges posed by over-population, environmental damage, technology-driven revolution, gross imbalances between rich and poor, and the onslaught of treatment-resistant diseases.

Yet despite this stirring introduction, there is no discussion of HIV/AIDS, or any other infectious disease, in the book. Indeed, the most common attempts to think through the real impact of these new types of threats to security come from science fiction, where a common theme is wholescale destruction through environmental catastrophe or the spread of infectious diseases. The fear of emerging or re-emerging diseases from the developing world insidiously attacking the rich world is an increasingly popular theme, moving from fiction and films, to such widely read science journalism as Laurie Garrett's *The Coming Plague* (Garrett, 1994).

For most of the twentieth century, the impact of two world wars and the succeeding 50 years of the Cold War meant that security remained defined almost entirely in military terms. The Maginot Line gave way to doctrines of nuclear deterrence, but in both cases it was assumed that national security rested upon conventional and nuclear military preparedness. With the end of the Cold War the short-lived hope for a 'peace dividend' was quickly replaced by recognition that the world was no safer, just more unpredictable. Increasingly wars have become civil conflicts, with escalating civilian casualties and sometimes belated attempts by international coalitions (as in the former Yugoslavia or East Timor) to impose an end to the carnage (Kaldor, 1999). At the same time, there is a growing recognition that the lines between military and non-military threats to security are being blurred, as in the case of terrorism and organized crime. It is difficult to perceive conventional military threats to either Western Europe or North America and increasingly their military power is likely to be involved in international policing activities.

Today's understandings of 'security' are increasingly characterized by a growing recognition of global interdependence. Collapse of civil order in one country can lead, for example, to mass population movements affecting other countries. Indeed, recent western European interventions in the Balkans and Africa have been fuelled by a fear of uncontrollable refugee flows as much as by humanitarian impulses. Similarly, the outbreak of an infectious disease in one part of the world can spread rapidly due to the volume and speed of travel. A nuclear accident such as the Chernobyl power station or environmental pollution (for example acid rain or the poisoning of waterways) can quickly impact on neighbouring countries and even farther afield. Some environmentalists speak of 'natural security' (basic access to food, health care, and a safe environment) (Hontiveros-Braquel, 1996) as a more useful term than 'national security' in an era when national boundaries are increasingly irrelevant as protection against many threats to human well-being and survival.

In one sense, the 'new' security threats are not new. The spread of the Black Death (bubonic plague) throughout Europe and Asia during the Middle Ages, or the movement of diseases such as syphilis and measles as part of the expansion of European empires into the rest of the world, are precursors to today's problems. What is new is both the instant transmission of news and information, and the growing acceptance that national sovereignty cannot be relied upon to respond to problems of global significance. The result is recent reports such as that compiled jointly by the Chemical and Biological Arms Control Institute and the Center for Strategic and International Studies in the US, which claims to 'directly link health and global security for the first time'. The report stresses the rapidity with which infections can spread; the threat of biological weapons; and the consequences for health of regional conflicts and failing states (Chemical and Biological Arms Control Institute, 2000). Indeed, there is evidence that officers in the Central Intelligence Agency (CIA) have been pushing their superiors to consider the impact of HIV/AIDS on national and global stability since 1990 (Gellman, 2000a).

Some of the more imaginative thinking about the meanings of security has come from the United Nations Development Programme (UNDP). UNDP has developed the concept of 'human security', locating security within a framework of 'the legitimate concerns of ordinary people', and encompassing safety from chronic threats of hunger, disease and repression, and protection from sudden disruptions in the patterns of everyday life. In brief, human security places emphasis on protecting individuals and communities, rather than state boundaries (Thomas, 2000, pp. 5–6). More recently, UNDP has linked these definitions of human security to the rapid changes and insecurities brought about through globalization: 'In the globalizing world of shrinking time, shrinking space and disappearing borders, people are confronting new threats to human security – sudden and hurtful disruptions in the pattern of daily life' (UNDP, 1999, p. 3). In some ways this conceptualization appears to echo the currently fashionable idea of a 'risk society', a term that is often used to characterize the contemporary condition. One analysis of 'human security' counterposes an emphasis on the community and individual, socioeconomic and environmental threats, and unstructured violence to the state-centred approach of more traditional concepts of security (Tow and Trood, 1999, p. 20).

There are strikingly different ways in which concern with new security threats can be approached. The concerns of the US government so far have largely been expressed as a desire to preserve stability and protect American citizens from risks originating from abroad. A quite different perspective begins with the conditions in the developing world that are creating such risks, including the rapid spread of infectious diseases, and argues for a greater global response to eliminate these. Speaking of the greater Mekong region – which straddles China, Burma, Thailand and Laos – Porter (1997, pp. 213–14) writes

> The nexus of HIV transmission across this territory is a metaphor for the globalisation of investment, trade and cultural identity. Although the dominant realist tradition in international relations studies conceives national territorial spaces as homogenous and exclusive, what is referred to as the 'new global cultural economy' has to be seen as a complex, overlapping, disjunctive order, which cannot be adequately understood in terms of centre–periphery, inner–outer, state border models of the past.
>
> Moreover, as Hurrell (1999, p. 259) points out, the new security threats: derive not from state power, military power and geopolitical ambition, but rather from state weakness and the absence of political legitimacy; from the failure of states to provide minimal conditions of public order within their borders; from the ways in which domestic instability and internal violence can spill into the international arena; and from the incapacity of weak states to form viable building blocks of a stable regional order, and to contribute towards the resolution of broader common purposes.

It is this latter perspective, of seeing HIV/AIDS from the standpoint of conditions in the developing world, that a fuller understanding of the links between globalization and health can be understood.

3.2   HIV/AIDS as exemplar of a new form of global security threat

As short as the history of the pandemic is, there already exists a genealogy of conceptual frameworks which have accompanied the global response to HIV/AIDS. The first attempts to create a framework for the control of HIV/AIDS grew out of the ideas of 'new public health' and the Ottawa Charter, with its emphasis on community participation and a move away from an overly medicalized view of health.[3] The first director of the Global Programme on AIDS (GPA), Jonathan Mann, was strongly influenced by the ideas around new public health and particularly the connection between human rights and vulnerability. Both human rights and support for non-governmental and community-based organizations were centrepieces in the policies developed under Mann. Simultaneously, stress was given to the links between HIV/AIDS and development, and the argument that the pandemic should be understood primarily as a development rather than a health issue. This was significant in the replacement of GPA, a WHO special programme, with UNAIDS, a co-sponsored programme of six UN organizations[4] (Mann and Kay, 1991; Tarantola, 1996). Whereas Mann stressed the link between vulnerability to infection and the lack of human rights, others have stressed the link between vulnerability and poverty, noting that widespread HIV/AIDS is often the result of mass human dislocations brought about by global economic forces (Ainsworth *et al.*, 1998; Barnett and Whiteside, 1999).

More recently, the enormity of the pandemic, and the need to focus political attention on its implications, has led to pressure to reconceptualize it as a political/security issue. Thus, a spokesperson for South African President Thabo Mbeki asserted, 'He has broken the tradition that seeks to make the disease just a health problem. HIV/AIDS is a socio-economic problem. It is a political problem that has reached the proportion of an international crisis. It threatens to destroy nations and continents' (Mankahlana, 2000). By 2001, when the UN General Assembly organized its Special Session on AIDS (UNGASS), this language had become commonplace as international rhetoric, if not yet of academic work.

Some work has already been done on the impact of globalization on the pandemic and indeed both the spread of, and the response to, HIV/AIDS fits well the spatial, temporal and cognitive dimensions of globalization discussed in Chapter 1. Much of the discussion to date has been in terms of the dislocations caused by rapid population movements, urbanization, the rapid spread of ideas and information, and the changes in government policies, often as a result of policies imposed by external organizations (Altman 1999b; Asthana, 1994; Ugalde and Jackson, 1995; Whiteside and Barnett, 1998). There is an irony in the World Bank putting increasing sums of money into HIV/AIDS work in countries such as Brazil and India where the Bank's own policies had helped weaken the health structures which might have helped prevent the spread of HIV. An important part of the impact of the pandemic, linked to social and economic upheavals, has been the increased vulnerability of women. Women are more likely to be unable to protect themselves against infection, to carry a greater share of the burden of care for those who are sick, and to have less access to treatments (Whelan, 1999). One of the most telling examples of how structural adjustment programmes affected the spread of HIV/AIDS is Kenya which shows a steep drop in attendance at STI (sexually transmitted infections) clinics following World Bank-enforced user charges for such visits (Epstein, 2001a, pp. 37–8; Lurie *et al.* 1995, pp. 539–46; Sen and Gurumuthy, 1998; Whitaker, 1997). In Zimbabwe some women's organizations argue that the obligation on doctors and counsellors to protect confidentiality further disempowers women and confirms existing sexual inequalities. These inequalities carry through into relative risks of HIV infection so that among Zimbabweans under 20 years of age, over 80 per cent of those infected are women (Musasa, 1998).

The most interesting link between HIV/AIDS and globalization, however, may be in the cognitive arena. Mobilization against diseases such as cholera or malaria involve the same sort of universalizing of western biomedical assumptions that are involved in programmes against HIV/AIDS. President Mbeki quite specifically invoked an anti-western critique in his questioning of the HIV hypothesis. The difficulties of involving traditional healers in AIDS campaigns (Steinglass, 2001), and the popularity of a series of apparent 'cures' for AIDS touted by various hucksters in India, Kenya and Thailand among others, underline the tensions between different world views which emerge around the spread of this deadly infectious disease. Equally HIV/AIDS has also seen appeals to concepts such as 'human rights' and 'good governance' as the basis of an effective response – sometimes as a condition for funding – as well as to particular forms of sexual and other identities. These contribute further to spreading particular western-derived understandings of the world. Thus, the principle of 'greater involvement of people living with AIDS' (GIPA), adopted with some fanfare at a ministerial summit in Paris in 1994, only makes sense in the context of western emphasis on the individual and his/her identity. Indeed the very notion of 'coming out' as positive is derived directly from the experience of the gay movement in western countries. In the same way, emphasis on community organizations of 'sex workers' or 'gay men' assume a particular form of identity politics which are by no means universally accepted (Altman, 2001, chapter 4).

Gradually the international community has come to recognize the broader political implications of HIV/AIDS. To take two examples, in late 1999 the UN Commission on Human Rights requested a Special Rapporteur to report on the effects of foreign debt on the 'full enjoyment of economic, social and cultural rights'. In the report released in March 2000, the three examples used were the HIV/AIDS epidemic in Africa, the impact of Hurricane Mitch on Honduras and Nicaragua, and the impact of debt relief on the 'worst forms of child labor convention' (UN Commission on Human Rights, 2000). A couple of months earlier the UN Security Council held a debate on the impact of HIV/AIDS in Africa, thus recognizing the implicit connection between security and the scale of the pandemic. At the meeting,

UNDP Executive Director Mark Malloch Brown (2000) pointed out:

> HIV/AIDS has a qualitatively different impact than a traditional health killer such as malaria. It rips across social structures, targeting a young continent's young people, particularly its girls; by cutting deep into all sectors of society it undermines vital economic growth – perhaps reducing future national GDP size in the region by a third over the next twenty years. And by putting huge additional demand on already weak, hard to access, public services it is setting up the terms of a desperate conflict over inadequate resources.

Already across large parts of southern and eastern Africa, over one-quarter of the adult population is infected with HIV, with rapidly increasing figures for other parts of the developing world including northeast India, Thailand, Cambodia, Papua New Guinea, parts of the Caribbean and some parts of the former Soviet Union. That HIV/AIDS predominantly affects young adults means that it hits the most productive members of society, with considerable social, economic and political consequences. Imagine the impact on a developing country that is losing large numbers of already scarce teachers, health workers, police and government officials.

Furthermore, the pandemic is inextricably connected with war and civil unrest. Wars fuel the pandemic as UNAIDS Director Peter Piot (2000) points out: 'War is the instrument of AIDS and rape is an instrument of war' (see also Epstein, 2001b, pp. 18–19). Summarizing the impact of war on the spread of HIV/AIDS, one Nigerian commentator identifies six factors as relevant: widespread rape by soldiers; massive and uncontrollable population movements; the creation of large refugee camps and the conditions for unprotected and forced sex within them; poverty leading to an increase in commercial sex; decline of literacy and access to basic prevention information; and the collapse of health services, leading to lesser ability to follow infection protection guidelines (Oyaku, 2001). All of these factors can be clearly identified in recent civil strife and warfare in the Congo, Sierra Leone, Rwanda, East Timor and elsewhere. Sierra Leone now has a growing HIV/AIDS epidemic, with estimates of widespread rape and an HIV-positive rate of almost 25 per cent among army recruits (Robinson, 2001).

There is almost certainly also close relationship between high HIV-infection rates and civil disorder. A report from the US National Intelligence Agency estimates that HIV prevalence among militia in Angola and the Congo was between 40 and 60 per cent, and earlier reports linked genocide in Rwanda during the late 1990s to widespread and deliberate HIV infection (Bertozzi, 1996, p. 174). In other parts of the world, such as Cambodia, Honduras and the former Yugoslavia, it is claimed that the presence of foreign troops has been a major factor in leading to the introduction of HIV/AIDS (Beyrer, 1998; Kane, 1993) and there is growing concern about the ways peacekeeping forces might help spread the disease. Given the high rate of casual sex engaged in by military personnel, HIV/AIDS has special implications. Military forces need to develop policies on how to best deal with the threat of large numbers of infections and the concomitant consequences this poses. Thailand began early in developing programmes to deal with HIV/AIDS within its military, but most countries have been slow to act, often extremely reluctant to confront the problem. This is despite growing concern about growing HIV-infection rates among military forces in Nigeria, Russia and Mozambique. Given the reality that the military is the primary source of political power and social order in many of the world's poorest countries, the existence of widespread prevalence of HIV-infection among military personnel threatens political and social stability more generally. It seems possible that the political hysteria to which President Mugabe appealed in Zimbabwe during the election campaign of 2000, and the violence directed against white settlers, was connected in various ways to the high rate of HIV-infection, not least as a way of denying the gravity of the epidemic.[5]

In summary, HIV/AIDS reduces life expectancy, distorts health budgets and creates a generation of orphans. It is already having an impact on the economic future of many countries because it attacks those who are most productive, and increases the burden of care on others. In many parts of sub-Saharan Africa, and potentially in other parts of the world, the loss of skilled workers and professionals is leading to a measurable decline in living standards, compounded by the growing burden on already under resourced health systems. In more and more villages, the care of children is falling upon grandparents, as the intermediate adult generation becomes sick and dies. In Zambia teachers are dying of HIV/AIDS faster than they can be replaced, and the World Bank believes that once HIV-infection reaches 8 per cent of adults, which is now the case in at least 20 countries, per capita growth declines by 0.4 per cent per year (*Financial Times*, 2000). The development of elaborate and expensive treatments for HIV/AIDS is merely increasing the gap between the epidemics of the rich minority and the poor majority and fuelling growing anger towards large pharmaceutical manufacturers.

3.3   The implications of how HIV/AIDS and security are perceived

Does it matter if HIV/AIDS is understood as a matter of security rather than, say, health or development? Yes, because how we conceptualize the pandemic will impact on the extent of political commitment governments bring to dealing with it. Defining HIV/AIDS as a health issue limits it to the province of one ministry, often without much political clout. Redefining the disease to encompass security issues almost inevitably pushes it higher on government agendas, making it a 'high politics' concern. This is now happening, as in the decision of the recently democratized Nigeria to bring HIV/AIDS under the jurisdiction of a committee headed by the President himself. Similarly, in Cote d'Ivoire a special ministry for HIV/AIDS has been created. Equally one of the reasons HIV/AIDS was relatively poorly dealt with in post-apartheid South Africa is that, under President Nelson Mandela, policymaking was confined to the Ministry of Health (Marais, 2000; Schneider and Stein, 2001).

In the lead up to UNGASS, there was considerable emphasis on the need for greater international resources to fight HIV/AIDS, centred on UN Secretary General Kofi Annan's call for a Global AIDS and Health Fund. Of course, more financial resources are needed and President Mbeki's call to link the struggle against HIV/AIDS to poverty is a powerful one. UNAIDS has been seeking to encourage debt-for-AIDS swaps with international lenders, with some success to date in Uganda, Zambia and Mozambique (UNAIDS, 2000a). More recently, the Harvard University economist Jeffrey Sachs has called for the US government to spend US$3 billion per year to buy HIV/AIDS drugs for African countries, a policy that would presumably also benefit US pharmaceutical companies. Indeed, it may be asked whether patients would benefit when many are in situations where compliance with complex treatment regimes would be extremely difficult. Sachs acknowledges that such a policy decision would need to be accompanied by a comprehensive programme of prevention and community care, along with a broader commitment to treat other diseases such as malaria and tuberculosis (Sachs, 2001).

One of the biggest problems in responding to the pandemic has been government timidity and denial. The Report of the UN Special Rapporteur cited above refers to 'the politics of indifference', blaming both rich and poor governments:

> Complicating further the fight against HIV/AIDS in Africa has been the cold-blooded abandonment of millions of poor Africans by their own Governments. Outside of South Africa, Uganda and Senegal, where enlightened political leaders have taken a proactive role to mobilise the population against the spread of HIV/AIDS, many African leaders have become obstacles to any concerted effort by donors, NGOs and civil society groups to contain the spread of the epidemic.[6]

Since the report was published in March 2000, other African countries, notably Botswana and Lesotho, have markedly increased their attention to HIV/AIDS.

The great irony is that we know how to prevent HIV transmission and it is neither technically difficult nor expensive. Most HIV transmission can be stopped by the widespread use of condoms and clean needles. Only in terms of blood screening and mother-to-child transmission does prevention involve the use of costly technology. But for this to be achieved requires major changes in behaviour, both individual and collective, which in turn require support for programmes that can infringe cherished religious and cultural beliefs. Effective HIV prevention requires governments to acknowledge a whole set of behaviours – drug use, 'promiscuity', homosexuality, commercial sex work – which they would often rather ignore, and a willingness to support, and indeed empower, groups practicing such behaviours. This requires a willingness to recognize that behaviours that are prohibited nonetheless exist and that the more it is stigmatized or even criminalized the more difficult it is to effectively prevent HIV transmission. In other cases, most particularly widespread forced sexual intercourse, it may mean an increase in policing activities. As Caldwell (2000) argues, 'The central plank in the victory over AIDS is the recognition by African governments of social and sexual reality. Millions of people are being allowed to die on the grounds that the only way they can be saved is by adopting a more "moral" way of life, indeed a way of life that is not their morality.'

Moreover, measures against HIV/AIDS may well involve changes to existing structures of power that threaten those who have most to lose. At a certain level, politicians understand that to speak of empowering women, abolishing stigma based on unpopular behaviours and status, can threaten the status quo from which they benefit. The development of peer education programmes or community health measures can, in turn, open up space for ideas of popular involvement in policymaking. This can lead to questioning of the allocation of government resources in ways which are potentially quite revolutionary. Where governments either depend upon or fear the power of organized religion, the major need for basic prevention information, centred around frank discussion of sexuality, is likely to be difficult. HIV/AIDS simultaneously requires new paradigms in understanding national and global security and state intervention in areas often regarded as both natural and private.

A few observers have seen the linking of HIV/AIDS to security concerns as leading to a new hypocrisy. Commenting on the UN Security Council debate (ACT UP Paris, 2000) saw it as:

> sinking even deeper into the 'prevention only' disaster ... to protect from the virus those who have not yet been contaminated, who will assure tomorrow the repayment of Debt as well as the stability of countries, even if it means to sacrifice those already sick, who are considered too expensive and already lost as labour; even if it means to give greater importance to prevention over access to treatments, to the epidemic over people with HIV.

The underlying concern, that in too many countries those who are already infected are largely disregarded, is justified. However, it is difficult to accept that this will be worsened by greater attention to the issue. Yes, governments and corporations are not motivated primarily by altruism. Yes, the primary concern is for the 'security' of the rich world, not for those already infected and those most at risk of infection. Yet if self-interest leads to a greater realism about HIV, it could not only improve prevention efforts, it may help break down stigma that remains a major part of the problem for those already infected. It may also remove a significant obstacle to people seeking even the limited assistance available in most parts of the world.

If questions of both national and human security cannot be separated from the larger context of global socioeconomic shifts, there are major implications for policy that go beyond specific HIV/AIDS programmes. For example, the spread of HIV/AIDS through needle use is, in turn, directly related to international (read US-directed) efforts to control the illicit trade in narcotics. It is well documented that such efforts have changed user habits in parts of southeast Asia, with a decline in opium smoking and a marked increase in the injecting of heroin (Asian Harm Reduction Network, 1998). Similar developments have taken place in Latin America, as a recent discussion of Mexico makes clear. Arguing that 'Mexico's current drug woes are an unanticipated side effect of the U.S. government's "successful" interdiction efforts in the Caribbean', Massing (2000) points out:

> Drugs are having devastating effects in Juarez (a border city). Five years ago drug addiction was virtually unknown there. Since then, the price of a dose of heroin has dropped from $100 to $3, and today thousands have become addicted. Hundreds of picadores, or shooting galleries, have sprung up to sell drugs...

Barnett and Whiteside (1999, p. 225) argue that there is a link between HIV/AIDS and the degree of social cohesion and overall level of wealth in a society. They write that the worst combination is low social cohesion and very unequal distribution of wealth. They also draw an important distinction between various forms of social cohesion, distinguishing that which may derive from civil society and that which stems from 'control through an authoritarian political or cultural system'. In the short run, it is possible that the latter will be more effective in controlling the spread of the pandemic. A good example is Cuba whose policies were criticized as unnecessarily restricting human rights, but nonetheless managed to limit the spread of HIV/AIDS through policies of compulsory quarantine. In the longer run, as is now evident in China and Vietnam, authoritarian governments can force risky behaviours underground, but are unlikely to be able to prevent them.

3.4 AIDS and the creation of 'global governance'

Perhaps the one bright note in what too often seems a tale of unmitigated failures and stupidity is the global solidarity that has been created in response to the HIV/AIDS pandemic.[7] The response, first by WHO which established the Global Programme on AIDS and then other UN organizations in creating UNAIDS, has been accompanied by the emergence of a whole set of local, regional and global community-based networks, with considerable links to multilateral and bilateral agencies (Altman, 1999a). The biennial International AIDS Conferences (the most recent was held in Durban in July 2000) generate considerable media coverage worldwide and have become significant symbolic expressions of global solidarity.

'Solidarity', of course, can easily become a meaningless shibboleth. Nonetheless, it is arguable that the HIV/AIDS pandemic has brought an extraordinary number of concrete examples, ranging from the individual commitment of someone like Paul Farmer (Kidder, 2000), to small groups such as 'buyers' clubs' in high-income countries that collect HIV treatments to send to people in low-income countries, to large companies donating resources, to multilateral organizations bringing the above together across countries. These forms of 'solidarity' can be seen as potentially filling the jurisdictional, participatory and incentive 'gaps' identified by Kaul *et al.* (1999) that result in shortfalls in global public goods such as infectious disease control. For all three types of gaps, there is a common need to reframe our understanding of the relationships between the national and the global, and among the state, individual citizens and community organizations.

It might be a useful exercise to ask how do we measure global solidarity. Possible measures might include the amount of development assistance, both bilateral and multilateral; contributions to peacekeeping forces; participation in international and regional events; compliance with international laws and agreements; and the role of nongovernmental activities. Ironically, while there has been a decline in official government aid, the past decade has seen a dramatic rise in global involvement by citizens and communities. This movement has, in turn, been legitimized through the mass community forums associated with UN conferences on women, population, human rights and the environment (Clark *et al.*, 1998; Desai, 1999; Smith and Pagnucco, 1998; Waterman, 1998).

This sort of interaction between citizens and governments can be seen as constituting what is increasingly described as a 'global civil society' which Falk (1999, p. 138) defines as 'the field of action and thought occupied by individual and collective citizen initiatives of a voluntary, non-profit character, both within states and transnationally'. As in other areas of global concern, HIV/AIDS has opened up possibilities for NGOs and indeed CBOs to influence and work with both governments and international agencies. From its inception UNAIDS sought to work in conjunction with CBOs and to avoid the formal and bureaucratic processes of the UN system. It created a place for 'NGO delegates' (not representatives) on its governing body, the Policy Coordinating Board, and sought to establish ways of consulting with all significant NGO networks to choose such delegates. Probably more important has been the steady recruitment of people from the community sector to work for UNAIDS, and a greater day-to-day engagement with HIV/AIDS organizations than is usual for international organizations.

Yet UNAIDS is caught in a contradiction it cannot resolve – its success depends on establishing co-operation among its UN co-sponsors within which territorial claims can be more important than policy outcomes. For UNAIDS to coordinate, say, the World Bank and UNESCO is rather akin to asking a sparrow to direct a herd of elephants. Increasingly, it has come to act as a focal point for global activities, providing information and contacts, and in some countries directly influencing policies by both donors and recipients. Such success within countries depends on its own representative establishing good relationships. Hence there are UNAIDS country officers in close to 50 countries, serving as the chair of the UN theme groups (comprising the UN co-sponsors of UNAIDS and the officers of the country's own AIDS bureaucracies).

Just as significant is the positioning of HIV/AIDS on the global agenda. Under Peter Piot, UNAIDS has been able to recapture some of the moral weight which Jonathan Mann gave to GPA. The inclusion of HIV/AIDS on the agenda of both the UN Security Council and Secretary General Annan's vision for the future of the UN may be its most significant achievement. The dilemma is that, while HIV/AIDS clearly cannot be contained within national borders, an effective response depends heavily on the will of national governments. However much international support and solidarity may be generated, governments have to be willing to allow effective prevention, counselling and treatment programmes which, in turn, require the use of community organizations to mobilize often marginalized and disempowered populations. The UNGASS meeting in 2001 revealed that many governments were uneasy about anything other than the most token involvement of PLWHAs and civil society organizations in policy debates. Enough countries also adhered to a conservative position to ensure the defeat of moves to mention men who have sex with men, drug users and sex workers among those who are most vulnerable to the disease.

But UNGASS did illustrate that the security concerns around HIV/AIDS have begun to affect governments, particularly those in sub-Saharan Africa. There is notable concern at the absence of senior political figures attending from countries in Asia and Eastern Europe with potentially huge epidemics. By declaring HIV/AIDS a 'global emergency' and creating a new Global AIDS Fund, the UN Special Session marked the final break from defining HIV/AIDS in purely health terms, thereby putting it on the national agendas of all governments. Commenting on the achievements of UNGASS, Piot stressed that 'It is our job to push the edges now' (Steinhauer, 2001).

If governments see HIV/AIDS as a matter of national survival, they are more likely to provide the degree of resources, political and financial, that are required. Indeed the paradox is that, while governments may often seem too weak to provide effective social programmes, they are usually strong enough to create extraordinary barriers against others who might provide them. A good example is Burma where the government presumably knows the potential threat of the pandemic but, for its own survival, places considerable obstacles in the way of domestic and global efforts to create effective HIV/AIDS programmes. Stronger governments, meaning governments that enjoy the support of their citizens, will feel less threatened by global efforts and be able, in turn, to create supportive environments within which international and nongovernmental efforts are most effective.

In the end it is only when governments in the most affected countries recognize that stopping the pandemic goes to the heart of national survival that one can expect the mobilization of will and resources to do what is needed. As long as only a handful of leaders in poor countries take the lead in making HIV/AIDS central to their agenda, and here President Mbeki for all his eccentricities stands out, the international community will be unable to respond effectively. The HIV/AIDS pandemic demonstrates only too tragically the need to put global health onto the table of 'high politics'.

Notes

1. I am grateful to the Macarthur Foundation for a Research and Writing Grant which provided support for the writing of this chapter.
2. For an early discussion see Thomas (1987).
3. The Ottawa Charter was a statement adopted by 38 countries in 1986 which identified good health as resting upon 'the empowerment of communities, their ownership and control of their own endeavours and destinies' (WHO/Health & Welfare Canada/Canadian Public Health Association, 1986).
4. The six co-sponsoring UN organizations were WHO, UNDP, UNICEF, UNESCO, UNFPA and the World Bank.
5. I write 'seems possible' quite deliberately as there appears to be a great reluctance to publicly discuss this hypothesis. Garrett (2000a) reports that she received death threats from ZANU thugs merely for asking about government failure to address HIV.
6. I should note that I do not necessarily endorse the particular allocation of blame – which is rather indiscriminate – on the part of the Special Rapporteur (UN Commission on Human Rights, 2000, para 40).
7. I do not use 'solidarity' here in the way it has been argued within human rights discourse. For example see Wellman 2000.

4
Far From the Maddening Cows: The Global Dimensions of BSE and vCJD

Kelley Lee and Preeti Patel

> The crisis is the price Europe has to pay for promoting, for years, an economic model in the food sector that favours intensive production and ignores basic standards of quality and food safety.
>
> (Maxwell, 1997)

> How food is exchanged across borders isn't very transparent or easy to understand.
>
> (Ricketts, 2000)

4.1 Introduction

During the past two decades, the public health community has come to realize that the world of infectious diseases is not one to be conquered once and for all by biomedical weapons, but a constantly changing one that continually throws up new challenges. At least 30 new disease-causing agents have been identified since 1973 including the human immunodeficiency virus (HIV), Hepatitis C, Nipah virus, Ebola virus, and *Borrelia burgdorferi* (cause of Lyme disease). At the same time, 20 well-known diseases previously thought to be effectively under public health control, at least in high-income countries, have increased in number of cases or geographical spread, in some cases in novel and potentially more virulent forms. These include tuberculosis, plague, diphtheria, malaria, yellow fever, dengue fever and cholera. Of particular concern is the development and spread of drug-resistant strains of such diseases as tuberculosis and malaria, as well as forms of bacteria resistant to antimicrobials (see Fidler in this volume). In 1998 infectious diseases accounted for up to one-third of the estimated 54 million deaths worldwide (US, 2000; WHO, 1998a).

A related source of public alarm in recent decades has been the succession of 'food scares', again most notably in high-income countries, where modern and increasingly globalized production methods have led to an unprecedented supply of food. In part, this preoccupation with food safety among the well-fed of the industrialized world is part of the so-called 'health transition', from a disease burden focused on infectious diseases, to more chronic and long-term health conditions. Nonetheless, public health alarm over such threats as *Escherichia coli*, salmonella and Listeria has meant that food safety has never before been under such close scrutiny. Foodborne illnesses have increased to unprecedented levels.[1] The global incidence of foodborne disease is difficult to estimate, but it has been reported that in 1998 alone 2.2 million people, including 1.8 million children, died from diarrhoeal diseases (WHO, 2000). These estimates are primarily related to microbiological problems as well as contamination (Brundtland, 2001).

The outbreak of bovine spongiform encephalopathy (BSE) in cattle in the UK from the 1980s, followed by human forms of the disease known as variant Creutzfeldt-Jakob disease (vCJD), bring together the twin threats of infectious disease and food safety. While the public health risks surrounding BSE/vCJD have received substantial national, regional and worldwide attention, the crisis has been largely portrayed as a UK problem requiring changes in British policies and practices to prevent further spread of the disease. The bans on the export of British beef and beef products, and restrictions imposed by many countries on blood donation by people resident in the UK for more than a given time period, have further characterized the parameters of the crisis along national boundaries.

This chapter argues, however, that the global dimensions of the BSE/vCJD crisis must be more fully explored. The study begins with a brief discussion of the crisis, including the detection of the first cases of BSE in the UK, government responses to growing public concern, and the reaction of the world community. It then explores how globalization has contributed to the emergence of the crisis by preventing a more timely response to the epidemic to protect public health and spreading the health risks far beyond the UK. It concludes that, without fundamental changes to the practices of the global food industry, human populations worldwide will remain vulnerable. This raises important implications for the making of public health policy and the strengthening of global governance for health.

4.2   The emergence of the BSE and vCJD crisis

Bovine spongiform encephalopathy (BSE) is the form of a group of diseases that affects a wide range of animals, known as transmissible spongiform encephalopathies (TSE), found in cattle. The transmissible agent associated with TSE is not fully understood, displaying many virus-like features (for example strain variation, mutation), but differing in its resistance to heat, ultraviolet and ionising radiation and chemical disinfectants. The prevailing belief is that TSEs are not caused by bacteria or viruses but a non-living agent, a special kind of 'prion' protein (PrP) which produces a slow biochemical reaction that modifies the protein molecules in the brain. The brain tissue affected takes on a characteristic sponge-like consistency as a result. Prions are unprecedented infectious pathogens that cause a group of invariably fatal neurodegenerative diseases by unique mechanisms. Prion diseases may present as genetic, infectious, or sporadic disorders, all of which involve modifications of the PrP. Prion proteins do not trigger a response from immune systems, making them impossible at present to detect. BSE, scrapie in sheep and CJD in humans are among the most notable prion-induced diseases (Prusiner, 1998).

In the UK, BSE was officially identified in November 1986 by the Ministry of Agriculture, Fisheries and Food (MAFF) as a newly recognized form of a neurological disease, although investigation of a cow in 1984 was subsequently shown to be infected with the disease. During the next two years to 31 December 1988, 2160 cases were confirmed in the country occurring on 1667 farms (Maxwell, 1997). British authorities conducted a survey to delineate what was common to the affected herds. Explanations for the initial appearance of the disease include the argument that the disease occurred spontaneously in cattle or that it entered the cattle food chain from the carcasses of sheep infected with a similar disease (scrapie). Whatever the original source, a common link was quickly discovered – those herds in which BSE occurred had been fed animal feeds containing certain tissues, notably brain and spinal cord. Such tissues are part of the by-products of slaughtered and dead animals (for example through disease) whose so-called mechanically-recovered meat (meat and bone meal, or MBM) is routinely incorporated into feed as a convenient and cheap source of protein. As some cattle became sick with BSE, most likely their carcasses were also recycled into animal feed and fed, in turn, back to other cattle. This explanation of the cause of BSE is supported by evidence that TSEs found in a variety of zoo and domestic animals such as gembok, nyala, kudu, eland, cats, Arabian oryx, Marmoset monkey, cheetah, ostrich, puma and civet, have their origin in animal feeds containing MBM derived from animal carcasses (Dealler, 1996, pp. 265–7).

By 1989, 400 new cases were appearing each week. By the early 1990s, BSE among cattle in Britain had turned into a major epidemic. Between 1986 and 1991 approximately 160,000 cases of the disease were confirmed and by 1992 100 new cases appeared daily. Within weeks of the identification of the first case of BSE, concerns were expressed about the human risk from eating meat from infected cattle. The British government's immediate response to the reported cases was to commission further investigation of feeding practices and transmissibility. This was later followed by a ban on the feeding of ruminant[2] feeds to cattle (June 1988), policy to slaughter all animals showing clinical symptoms, compensation to farmers at 50 per cent for confirmed cases and 100 per cent for negative cases (August 1988), prohibition on the use of milk from suspected animals other than for feeding a cow's own calf (December 1988), ban on use of certain specified bovine offals for human consumption (November 1989) followed by a European ban, full compensation for farmers up to a ceiling (February 1990), and formation of the Spongiform Encephalopathy Advisory Committee (SEAC) by MAFF and the Department of Health in April 1990 (Brown *et al.*, 2001).

Throughout this period MAFF officials continued to deny that there was any risk to human health and that consumers should continue to use beef products. Furthermore, it was argued that, with the above precautionary measures in place, notably measures to prevent infected meat from entering the human food chain, any potential and hypothetical risk would be avoided. In the meantime, further research was commissioned on the infective agent and its transmissibility, both maternally (from mother to calf) and horizontally (from one cow to another). Stricter guidelines were adopted for rendering plants and abattoirs although, as has since been reported, were not supported by sufficient resources to ensure effective enforcement (BSE Inquiry Report, 2000).

Despite these efforts, reported cases of BSE continued to rise, surpassing government estimates and peaking in 1993 at 0.3 per cent of the British herd. Public disquiet increased and in April 1990 education authorities began to withdraw beef from school meals. Even more damaging for the government's position was a ban by 15 countries, including the US, on British beef imports. Deeply concerned at the economic impact of such bans, MAFF continued to downplay the risks to human health. With so little known about the disease at the time, scientists did not have the empirical evidence to definitively challenge this stance. For the most part, they advised that the risk of cross-species transmission of prion disease by the oral route, using material from an animal that was not yet overtly sick, was low and that, after the specified offals (especially the brain and spinal cord) had been removed from the human food chain, the risk was lessened further (Ridley and Baker, 1998, p. 190). Yet tensions between MAFF and public health officials were soon to develop.

The government's official position on the risk posed by BSE to human health changed dramatically in March 1996 when SEAC announced that ten human cases of a new and previously unreported form of Creutzfeldt-Jakob disease (CJD) had been found. CJD is a fatal disease of the brain of middle or early old age that causes a spongy deterioration of the brain and rapid progressive dementia. The disease is usually fatal within one year. The new variant of CJD was different in the relatively young age of death (all under 42 years) and the short duration of the illness from onset to death. While no scientific evidence thus far linked vCJD to BSE, its similar characteristics to

TSE in other animals led to the hypothesis that the human cases might be associated with the consumption of BSE-infected beef. This conclusion was subsequently supported by a WHO Consultation on Public Health Issues related to Human and Animal TSE in April 1996 (WHO, 1996).

Reaction in the UK and abroad to this conclusion was rapid and substantial. A Beef (Emergency Control) Order (SI 1996 No. 961) was adopted on 29 March 1996 that prohibits the sale for human consumption of any meat from bovine animals over a certain age. Amendments to the order over the following weeks introduced the use of identification documents. A flurry of amendments to the BSE Order were also adopted which concerned the feeding of ruminant feeds to other farm animals (that is pigs, sheep, chickens) and compensation levels for farmers. Finally, regulations and orders were quickly passed relating to abbatoir practices, fertilizer production and culling of potentially infected herds.

Abroad, the European Union announced a prohibition (Commission Decision 96/293) on 27 March 1996 on the export from the UK of live cattle; meat products destined for the human or animal food-chains; semen and embryos; bovine materials destined for use in medicinal products, cosmetics or pharmaceutical products; and mammalian derived meat and bone meal. Other countries around the world soon followed suit. The British government responded by banning the sale of meat from all cattle over 30 months old, and culling 60,000 cattle over that age between May–July 1996.

The full human consequences of BSE are still unknown. As of July 2002, there have been 115 confirmed cases of vCJD in the UK. Estimates of the ultimate size of the epidemic vary enormously, from one thousand to 10 million potential cases, although it is agreed that the number of new cases is gathering pace, increasing 20 to 30 per cent per year (Cookson, 2001). The main difficulty is the lack of information – how many infected cattle entered the food chain, what types and quality of meat were used for which food products (Meikle, 2001), what dose of infective agent is needed to cause disease in humans, how many people are presently incubating the disease and how long is the incubation period? And it is these uncertainties that make the global dimensions of BSE and vCJD so worrying.

4.3   The global dimensions of BSE and vCJD

The BSE crisis has widespread implications for human and animal health around the world. While the issue continues to be seen as largely a British problem, growing evidence suggests that the global trade in potentially infected animal feed, live animals, beef and beef products, and biologicals may have spread the disease to other countries. More fundamentally, a state-centric analysis of, and policy response to, the BSE crisis may overlook the role of certain food production practices in contributing to the initial outbreak of the disease including pressures for low-cost and high-volume food and the primacy given to commercial over public health interests. In this sense, the lessons from BSE have direct relevance for public policy far beyond a single country.

Despite the sometimes frantic efforts by public officials in other countries to reassure their constituencies that a similar crisis could not happen elsewhere, the transborder nature of many of the risk factors behind the crisis suggest otherwise. By 1996 cases of BSE had been reported in ten countries and areas outside of the UK. In France, Portugal, Ireland and Switzerland, the disease occurred in native cattle as a result, it is believed, of the importation of infected animal feeds from the UK. In Canada, Denmark, the Falkland Islands, Germany, Italy and Oman, the cases occurred only in cattle imported from the UK (see Table 4.1).

As well as the incidence of BSE in cattle worldwide, there have been a small but significant number of vCJD cases. In Britain the onset of illness in the first case of vCJD occurred in early 1994, nearly a decade after the first case of BSE was recognized in cattle (Brown *et al.*, 2001). Through the end of November 2000, the overall tally was 87 definite or probable cases of vCJD in the UK, three in France and one in the Republic of Ireland. While the Irish patient had lived for some years in England, none of the French patients had lived in or visited the UK and so their infection must have

Table 4.1   Reported cases of BSE in the United Kingdom and other cases (as of December 2000)

Country	Native cases	Imported cases	Total cases
United Kingdom	180 376	0	180 376
Republic of Ireland	487	12	499
Portugal	446	6	452
Switzerland	363	0	363
France	150	1	151
Belgium	18	0	18
Netherlands	6	0	6
Liechtenstein	2	0	2
Denmark	1	1	2
Luxembourg	1	0	1
Germany	3	6	9
Oman	0	2	2
Italy	0	2	2
Spain	0	2	2
Canada	0	1	1
Falklands	0	1	1
Azores (Portugal)	0	1	1

Source: Brown P. *et al.* (2001), 'Bovine Spongiform Encephalopathy and Variant Creutzfeldt-Jakob Disease: Background, Evolution, and Current Concerns', *Emerging Infectious Diseases*, 7(1), January–February: pp. 6–16.

come either from beef or beef products imported from the UK (comprising about 5–10 per cent of the beef consumed in France) or BSE-affected cattle in France (Brown *et al.*, 2001).[3]

Once again, limited knowledge makes it difficult to predict with accuracy the eventual scale of the public health impact of BSE/vCJD in other countries. Furthermore, the sensitive political and economic risks, including potentially adverse impacts on the farming and food industries, can be a disincentive to reporting the disease. Nonetheless, it is clear that better understanding of the global dimensions of BSE/vCJD would be a critical part of the eventual picture. Three aspects are now considered – the globalization of the food industry, the trade in animal feeds and the export of other potentially infected products such as blood, plasma and organs.

4.4   The globalization of the food industry

The food industry is today dominated by ten multinational corporations (MNCs) that dominate throughout the food production chain – agricultural production, storage, processing and packaging, transporting and marketing (Lang, 1999). Four companies control 90 per cent of the world's exports of corn, wheat, coffee, tea, pineapples and tobacco. The current phase of globalization is characterized by increasing concentration at the national, regional and international levels. The UK food industry, for instance, is one of the most concentrated in Europe and the industrialized world. In 1995, three companies (Unilever, Cadbury Schweppes and Associated British Foods) accounted for two-thirds of total capitalization in UK food manufacturing (Lang, 1999). Nineteen of the top 50 Europeans food companies are British and UK companies are second only to those of the US in the level of foreign direct investment in other companies.

The increasingly global nature of the food industry poses changing opportunities for, and risks to, public health. In terms of risks, the more integrated nature of the food production process, from plough to plate, has been accompanied by an increased transborder incidence of foodborne diseases. Such diseases range from the relatively familiar *salmonellosis*, *E coli*, listeriosis, botulism and *campylobacteriosis*, to the even more serious conditions of cholera, typhoid and poliomyelitis. WHO estimates that foodborne illness is one of the most widespread health problems in the world, but reliable statistics on incidence, particularly in the developing world, are hard to obtain (Wilson, 2001). People suffering from a foodborne illness often do not know the cause and do not, therefore, seek medical care. Even if victims go to a doctor, few countries have adequate health care systems in place to effectively monitor outbreaks. Global surveillance is hindered by weaknesses in infrastructure and the relatively low priority given to foodborne diseases by public health authorities. It is estimated that less than 10 per cent of actual cases are reported (Motarjemi and Kafterstein, 1997). Global control is further hindered by the 'fast tracking' of trade agreements which are currently overwhelming national systems of food inspection and regulation (Public Citizens Global Trade Watch, 1997).

No one is immune from the impact of the immense changes in how food is grown, processed, distributed, marketed and sold around the world. The transmission of BSE to human populations via the consumption of contaminated beef raises questions over the extent to which potentially infected food products have been consumed worldwide. The food industry in the UK, as in other high-income countries, mirrors the global structure of the industry. The food industry is the largest manufacturing industry led by such companies as Unilever, Grand Metropolitan and Hillsdown. Competition by food retailers is also fierce, resulting in pressures all the way down the production chain to reduce costs. Until the BSE crisis, Britain was a major exporter of beef and beef products. The meat industry was worth £10 billion (US$15 billion) annually, with beef accounting for £4 billion (US$6 billion) (Workers Inquiry, 1998). Industry interests are protected foremost by MAFF, but also by the Department of Trade and Industry.

The original source and movement of animal and animal products can be masked by trade data that often do not record the processing and re-export of products. Many products such as sweets and puddings, not usually thought of as containing 'meat', may contain beef-derived ingredients (for example gelatin). Although the amount of beef-derived material in such products is likely to be extremely small, if the infectious agent is separated out in such a way as to find its way into these products, it may be assumed that a large proportion of the population can be exposed to an extremely low level of infectivity (Ridley and Baker, 1998, p. 197). The problem of illegal trade in meat is also a major concern. Consequently, public health officials should be aware of the potential spread of infected material to other parts of the human food chain as a consequence of standard production practices and trade patterns. What is considered 'safe' to eat thus becomes far less clear, raising a range of policy questions: is enough being done to protect human populations from BSE? How can the spread of BSE to other countries be prevented? What can scientists do to improve communication about the risks from BSE? (WHO/FAO/OIE, 2001)

4.5   The global nature of the animal feed industry

A key feature of the BSE crisis is the trade in animal feed. Most artificial feeds are cereal-based with the addition of vitamins, minerals, protein supplements and drugs (for example antibiotics). The practice of adding protein-rich pellets made from meat and bone meal (MBM) to animal feeds grew from a post-World War II strategy of increasing the milk yield of dairy herds (Dumble, 2001). Pellets are produced by rendering plants that take the inedible parts of animal carcasses and, using chemical substances, mechanically produce two products: tallow (fat) for use in soap, lipstick and cosmetics and protein-rich material (MBM). A wide variety of animals are disposed of through rendering plants including domestic pets.

Around the late 1960s to early 1970s, bone meal (known as protein concentrates) began to be added to animal feeds as an economical source of nutrition. Such products have been widely used to feed pigs, chickens, sheep and other animals destined for human consumption. Feed compounds are produced by a small number of countries and exported worldwide. In 1987 there were about 40 rendering plants in the UK processing 1.3 million tonnes of raw material (Khor, 1996). In 1989, Britain exported 25,000 tons of potentially infectious MBM to Europe (Dealler, 2001). The main export markets for British animal feeds were France, Germany, Denmark and Switzerland in Europe, Dubai and other countries in the Middle East, and Africa. Much of this, in turn, can be re-exported farther afield. Of the three million tons of MBM produced by the European Union, an estimated 500,000 tons have been re-exported, mostly to Eastern Europe, Asia and the US (Olson, 2000). Disturbingly, following the ban on ruminant-based feeds in the UK in 1988, many suppliers began to export legally or otherwise surplus stock that could no longer be sold in Britain. Many of these supplies went to other European countries, some of which have since experienced cases of BSE. Countries where cases of BSE traceable to imported cattle or animal feeds from the UK include Oman, Denmark, Germany, Portugal, Falkland Islands, Ireland, France and Canada. BSE has not occurred in the US or other countries that have historically imported little or no live cattle, beef products, or livestock nutritional supplements from the UK (Brown *et al.*, 2001). However, it is notable that ruminant-based feeds continue to be used to feed a range of animals in many countries, including the UK and US.

Detailed research on the animal feed industry, the safety of its ingredients and the transnational nature of its activities is urgently needed to clarify the potential public health risks. This should begin with a questioning of the economically driven practice of feeding animal protein to herbivores, a widespread practice that continues today in many countries. Measures aimed at minimizing the risk of infected animals being used for such purposes are not comforting given that so much is unknown about TSEs. The unknown incubation period of TSEs, in particular, raises the possibility that animals are slaughtered and consumed before disease manifests itself. Finally, the lack of data on how feeds are produced, and their export throughout the world, needs to be addressed as a preliminary to understanding any future outbreaks.

4.6 The global export of blood and other biological products

The consumption of beef and beef products are not the only unknown source of risk to public health. In July 1998 concerns began to be raised over the possible transmission of vCJD through blood products and other biologicals. It is estimated that 80,000 blood donors in the UK could be carrying vCJD and that one in 125 patients given a transfusion could receive contaminated blood. In February 1998 SEAC recommended that British blood not be used in the manufacture of plasma products because of the possibility of infection. From September 1998 it was agreed that all plasma would be purchased from the US where vCJD is yet unknown (Murray, 1998). However, this has proven more difficult than anticipated given worldwide shortages in plasma (Warden, 1998).

The global blood industry, worth an estimated US$5 billion annually, has proven fallible in the past. As a consequence of unregulated private collecting and processing, most worryingly from individuals in low-income countries paid for their donations, widespread infection through blood contaminated with HIV/AIDS and hepatitis B has occurred. Poor or nonexistent screening of blood donors and donations, the pooling of donations into large batches, the non-use or lack of testing facilities and the worldwide trade in blood and blood products raise serious public health concerns (Starr, 1997).

In relation to vCJD, research has shown that CJD can be transmitted to humans by treatment with natural human growth hormone or the grafting of tissues surrounding brain tissue (WHO, 1996). Two neuroscientists, Laura and (the late) Eli Manuelides of Yale University, had previously shown that injections of human blood, like injections of brain material taken from kuru and CJD victims, transmitted the disease across the species barrier to laboratory animals (Dumble, 1997). This implied that blood was the vehicle that carried the agent of CJD around the body until it began to reside in the brain. This also meant that the blood route was the key route to the transmission of CJD from a primary to a secondary host. As distinct from infections such as influenza, but in common with AIDS and hepatitis B, this indicated that recipients exposed to human pituitary gland hormone injections, or to blood or organ transplants from a donor with CJD, risked becoming secondary CJD hosts once contagious material entered their bloodstream.

The concerns over vCJD and blood were confirmed when it was reported in early 2001 that blood products donated by three people who were later struck down with vCJD have been sold to 11 countries (Meikle and Bellos, 2001). Changes in surgical practices in the UK and elsewhere began to be considered, including the use of disposable instruments in selected procedures to minimize the risk of infection.

4.7   The global governance of food safety

The issues raised by the BSE/vCJD crisis and, in particular, the global dimensions of the crisis, raise important implications for the making of public health policy. Historically, responsibility for public health has been firmly located at the national level and within the public sphere, notably a ministry of health that has relatively weak powers compared with other government departments. Measures adopted to control the epidemic have been largely focused on the UK – culling of herds, tagging and tracking of individual cattle and regulation of the livestock industry. More recently, outbreaks of the disease in other countries and the European Union have led to the adoption of safety measures but to a far lesser extent than in the UK. Given the global dimensions of the BSE crisis described above, to what extent are national and to a lesser degree regional public health measures alone sufficient to address what are effectively transborder health risks?

One of the results of the BSE/vCJD crisis has been a fundamental shaking of public confidence in food safety and the government authorities responsible for ensuring this. In the UK, the BSE Inquiry was initiated by the Labour government to review events prior to and during the crisis, and the policy-making process surrounding the public health response. The inquiry was also an attempt to restore public confidence in the government's handling of the issue. Importantly, issues of governance have been central to the inquiry. The lack of transparency in decisionmaking, primacy given to the representation of industry interests through MAFF, the inappropriateness of government assurances amid scientific uncertainty and weaknesses in the representation of public health interests have all become apparent through inquiry proceedings. The creation of the Food Standards Agency, with a broad public health and food safety remit, has been a result of the review.

The lessons to be learned from the BSE/vCJD crisis for global health focus on the need for a better balance among all interests concerned with food production and consumption. An important feature of the BSE Inquiry has been open access by the public to proceedings, the opportunity to participate in its discussions and, notably, the establishment of a website offering transcripts, lists of committee members and biographies, schedules of events and background documents.[4] The constellation of interests in the BSE crisis, however, are global as well as national and regulation of the food industry, in particular, is highly variable from country to country. Given the global nature and scale of the industry, is there a need for greater standardization of food regulation across countries and coordinated at the global level? The food industry is well represented at the global level through, for example, the International Chamber of Commerce. In contrast, how are public health interests represented at the global level? Neither WHO nor the Food and Agriculture Organization (for example Codex Alimentarius) play such a role.[5] One of the challenges for creating a more effective system of global governance for health, therefore, will be wider participation in key forums of decision-making (Lee, 1998b).

Recent developments do not bode well for the democratization of global health as food safety continues to take second place to industry interests. The apparent precedence given to food industry interests over public health concerns is illustrated by the 1997 decision of the World Trade Organization (WTO) to reject the European Union's ban on hormone-treated beef imported from the US on the grounds that there is insufficient scientific evidence of a public health risk. The ban was enacted in 1988 because of concerns that beef, routinely treated with five hormones (for example testoterone, progesterone) to encourage growth and lactation, may be a danger to human health. The trade is worth an estimated US$250 million or 10 per cent of America's total beef exports (Andrews, 1997).

The significance of the WTO ruling is the precedent it sets for challenging other 'trade barriers' based on health concerns. The decision places the burden of proof on public health officials, creating a 'wait and see' process for sufficient evidence to be presented of actual harm. Undoubtedly risk assessment is a difficult task, made particularly so by inadequate or incomplete scientific information. The weighing up of costs and benefits is an imprecise exercise in such cases. Nonetheless, the example of BSE/vCJD shows that the economic benefits, in terms of employment, profits and taxation revenue, continue to be given priority over potential and future human and environmental costs. The latter derives from the resilience of the infective agent to heat and chemical treatments, making it unclear how long it can remain active. In the UK, fears of the financial cost of having to destroy entire herds because of an infected cow led to some farmers burying carcasses underground. Environmental groups have raised concerns over the possible seeping of the infective agent into ground water and eventually further afield. The widespread use of meat and bone meal for horticultural purposes is also seen as creating unknown long-term environmental impact. Is such a 'wait and see' approach to public health risk, as opposed to a more precautionary approach, acceptable when the potential impacts can be widespread and irreversible?

The current debate over GMOs is again raising similar issues regarding the prioritization of large multinational business interests over concerns for public health. Government policy on the current debate in Europe, rightly or wrongly affected by public sensitivities to the BSE crisis, appears to confirm the 'wait and see' approach. While the mass media have admittedly sensationalized many aspects of the issue (dubbing them Frankenstein foods), underlying issues surrounding the nature of public policy debate remain important to recognize. Government and industry assurances of the safety of genetically-modified foods have so far been based on limited and short-term research that focuses on the rewards of high-yield, pest-resistant and less perishable crops. Importantly, longer-term and non-economic costs to human health and the wider environment, both in the UK and worldwide have received far less attention. Cross pollenization of wild species, impacts on the natural foodchain (for example insect pests eaten by birds), monocropping and carcinogenic effects remain unknown. Significantly, unlike the BSE crisis, public opinion has been swift to mobilize and articulate its opposition

(McCarthy, 1998), raising questions about the nature of the global food industry.

4.8   Conclusions

There are complex and multi-factorial reasons for the resurgence of many infectious diseases. Infectious diseases remain a major public health problem in low-income countries, although their impact in the industrialized world has been substantially reduced since the twentieth century by mass immunization and the use of antimicrobials. However, continued progress is challenged in the face of emerging pathogens (for example HIV/AIDS, vCJD, new influenza strains) and forms of diseases such as tuberculosis that are resistant to existing treatments or antimicrobials (Donnelly and Ferguson, 1999, p. 1). Furthermore, the epidemiology of many diseases has been affected by global processes of change (for example climate change, patterns of human migration, global economic and political instability, terrorism) that go far beyond the health sector. These forces create major challenges for health systems in all countries in the form of transborder health risks that do not respect national boundaries. These challenges have encouraged a shift in perspective from *international* to *global* health co-operation (Lee, 1998b).

This chapter argues that the BSE/vCJD crisis goes far beyond the UK and can be seen as a potentially global health problem. The crisis illuminates weaknesses in the existing system of governance around public health issues and, in particular, the primacy of increasingly globalized corporate interests over the protection and promotion of public health. The UK has been blamed by many for the spread of BSE and vCJD to continental Europe and further afield (Jutzi, 2001). However, state-centric responses to a crisis brought about by practices with global dimensions is unlikely to be effective in the longer-term.

Since 1985, when the then mystery disease now known as BSE emerged in Daisy a dairy cow from Kent, the total number of BSE-infected cattle has continued to grow. By 2001, the UK had 181,284 cases of BSE. Although the rate of increase has slowed decidedly, peaking in 1993, the last cases will probably not be before 2006. Importantly, there are a growing number of cases in other countries. BSE is now increasing rapidly on continental Europe with, for instance, seven cases in Germany in 2000 and 94 in 2001 (Dealler, 2001), and the first case in Japan was reported in September 2001. Perhaps even more worryingly, there are reports of vCJD-like illnesses outside of Europe including South Africa, Pakistan and India. Some believe that this is the result of the sale of attractively priced BSE-suspect meat and protein-rich animal pellets to low-income countries, raising fears that the human and environmental consequences worldwide will ultimately dwarf the European crisis.

Notes

1. Foodborne illnesses are defined as diseases, usually either infectious or toxic in nature, caused by agents that enter the body through the ingestion of food. In industrialized countries, the percentage of people suffering from foodborne diseases each year has been reported to be up to 30 per cent. In the US, for example, around 76 million cases of foodborne diseases, resulting in 325,000 hospitalizations and 5000 deaths, are estimated to occur each year (WHO, 2000).
2. Ruminants are any *artiodactyl* mammal (placental mammals with an even number of toes) of the suborder *Ruminantia*, the members of which chew the cud and have a stomach of four compartments. The group includes deer, antelopes, cattle, sheep and goats, as well as other animals that chew the cud such as camels.
3. From a European standpoint, it would be more troubling if imported beef were the source as most European countries import beef or beef products from the UK although in smaller quantities.
4. See www.bseinquiry.gov.uk/index.htm.
5. In partnership with the Food and Agriculture Organization of the United Nations (FAO), WHO provides for the Secretariat of the Codex Alimentarius Commission (CAC). The standards, guidelines and recommendations of the CAC are referred to in the World Trade Organization's Agreement on the Application of Sanitary and Phytosanitary Measures (SPS Agreement) as the international benchmarks for national requirements and thus for international harmonization.

5
Think Global, Smoke Local: Transnational Tobacco Companies and Cognitive Globalization

Jeff Collin

5.1 Introduction

The magnitude of the global tobacco epidemic is becoming increasingly familiar. While an estimated four million people now die annually from tobacco related diseases this figure is expected to reach ten million by 2030, by which time over 70 per cent of such deaths will occur in low-income countries (WHO, 1999a). There is also increasing awareness of how mortality and morbidity have been exacerbated by the behaviour of transnational tobacco companies (TTCs). The globalization of the tobacco industry has proceeded to the extent that 75 per cent of the world cigarette market is now controlled by just four companies: Philip Morris, British American Tobacco (BAT), Japan Tobacco/R.J. Reynolds and the China National Tobacco Corporation (Crescenti, 1999). The latter's share can be solely attributed to its dominance of the enormous Chinese market, but the remainder have been assiduous in their pursuit of growth through worldwide expansion. Each company now owns or leases plants in at least 50 countries throughout the world (Hammond, 1998). Philip Morris saw its global revenues increase by 226 per cent to US$27.4 billion between 1989 and 1999, and BAT now sells 900 billion cigarettes per year in 180 countries (Maguire, 2000). Seventy per cent of these are sold in Africa, Asia, Latin America and Eastern Europe (Saloojee and Dagli, 2000).

In common with most assessments of the diverse health impacts associated with global change, and perhaps unsurprisingly given the culpability of such major actors, much attention has focused on the economic aspects of the relationships between globalization and the tobacco epidemic. There has been great interest in issues such as the consequences of trade liberalization for tobacco consumption (Bettcher and Shapiro, 2001; Callard *et al.*, 2001; Taylor *et al.*, 2000), the involvement of tobacco companies in smuggling

(Centre for Public Integrity, 2000; Joossens and Raw, 1998; Joossens *et al.*, 2000) and the national economic consequences of trends in tobacco production, consumption and control (Chaloupka *et al.*, 2000; Jha and Chaloupka, 1999; Lightwood *et al.*, 2000). Such themes clearly constitute a necessary part of understanding the globalization of the tobacco industry.

A full understanding must also, however, consider what Lee (2000) terms the cognitive dimension of globalization:

> The cognitive dimension of globalisation concerns changes to the creation and exchange of knowledge, ideas, norms, beliefs, values, cultural identities and other thought processes. How we think about ourselves and the world around us is being changed by globalisation. The causes of this are varied including the mass media, educational institutions, think tanks, scientists, consultancy firms and 'spin doctors'.

Within the burgeoning literature addressing global impacts on health, this remains very much a neglected dimension (Lee, 2000), an omission that doubtless reflects the intrinsic difficulties in studying such comparatively abstract and often non-quantifiable phenomena.

This chapter examines the ways in which TTCs have sought to shape 'our mental frameworks' (Gill, 1995) as an integral part of their strategy to secure rapid sales growth beyond their traditional markets in North America and western Europe. In particular, it illustrates the attempts of TTCs to establish a global presence for, and awareness of, their key brands; documents outline strategies employed to associate cigarettes with modernity, prosperity and internationalism in its creation of the 'global smoker'; and discusses the far-reaching efforts of the industry to structure the discourse within which tobacco issues are understood and debated worldwide. The chapter concludes by indicating ways by which characteristics of cognitive globalization can be employed to advance tobacco control.

The tobacco industry is perhaps uniquely amenable for such a study as a result of the disclosure of large numbers of internal documents, primarily due to litigation in the state of Minnesota (Ciresi *et al.*, 1999). As part of the Minnesota Settlement Agreement, major tobacco companies[1] were required to place around 40 million pages in a publicly accessible depository in Minneapolis, while BAT was obliged to create a similar depository of around 8 million pages in Guildford, England.[2] The latter provides a particularly rich resource, reflecting the dispersal of BAT's trading operations and the greater candour of the company's documentation, and it is analysis of these documents on which much of the chapter is based. The industry documents provide a powerful insight into the operations of the TTCs, allowing a mapping of the tobacco industry that has been likened to the human genome project in its implications for global public health (Glantz, 2000).

5.2 The promotion of global cigarette brands

> Logos, by the force of ubiquity, have become the closest thing we have to an international language, recognised and understood in many more places than English. (Klein 2000)

The most striking symbol of the dramatic recent expansion of the TTCs in low- and middle-income countries has been the rise of the global brand. Labels such as Lucky Strike, Winston, Camel and above all Marlboro have emerged as emblematic of the multiple guises of the globalization of the tobacco industry, the ubiquity of such logos serving as both cultural signifiers and economic engines. Indeed the concept and role of the international brand is intrinsically linked with the globalization of the tobacco industry:

> International brands are defined as those brands that are available in a number of markets and currently sell, or have the potential to sell, significant volumes in the future. They are generally priced at a higher or premium level, have consistent pack designs and communications to the smoker with a clear target consumer in mind.... The fact that a 'foreign' brand is sold on another market is not sufficient to justify its description as an International Brand, because the latter involves a mix of global availability, plus perception of internationality to the consumer. (BATCO Marketing Intelligence Department, 1994)

Such brands constitute a rapidly growing segment of the world cigarette market that is now approaching one quarter of total annual sales (BAT, 2001b), with annual sales growing about 5 per cent during the 1990s (Herter, 2000). More important than their significance in terms of volume, however, is the greater profitability of these brands given higher prices, production volumes and economies of scale that explain their strategic centrality. BAT's merger with Rothmans in 1999 increased its share of this key segment from 11.3 per cent to 17.6 per cent, improving operating margins from 16.8 per cent to 18.1 per cent (BAT, 2000a). Their promotion has been the engine of business expansion, with company strategies built around identifying and supporting a handful of 'drive' brands.

The exemplar of this approach is the success achieved by Philip Morris through Marlboro, transformed from an American brand with stagnant sales in the early 1960s, to a truly transnational business phenomenon and 'one of the quintessential global brands' (Klein, 2000). Despite the generalized growth of premium global brands, Marlboro continues to dwarf its competitors, accounting for 8.4 per cent of global cigarette consumption (Hammond, 1998). In 1992 a review of the commercial value of product trademarks in the magazine *Financial World* placed Marlboro first, comfortably ahead of

Coca-Cola, Budweiser and Nescafe (Ourusoff, 1992). A more recent review of the top global brands placed Marlboro eleventh, although it would now arguably be second only to Coca-Cola if the distinction between corporate and product brands drawn by *Financial World* was maintained (*Business Week*, 2001). What makes Marlboro such an interesting case in the context of cognitive globalization is the means by which such pre-eminence was achieved:

> [I]t seems safe to say that one reason why Marlboro, valued at US$31 billion, towers over competing tobacco brands is that Philip Morris has underwritten the brand with much bigger advertising expenditures (US$118 million). In contrast RJR spent US$18.7 million on Winston and US$69 million on Camel. (Ourusoff, 1992)

The global rise of Marlboro is inextricably linked to the success of its advertising and marketing. The 'Marlboro Country' campaign became one of the most successful and enduring promotions ever developed, based on powerful American imagery of a strong independent cowboy within an immense rugged landscape. The Marlboro Man was declared by *Advertising Age* to be the number one advertising icon of the twentieth century (Yach and Bettcher, 2000), and the campaign transformed the fortunes of the brand. It led Marlboro to global dominance via the insistent and ubiquitous message that he 'liked his coffee strong, his horse powerful and fast, and his smoke with a full-bodied taste – to hell with its tar and nicotine content or the harping of the sissy Surgeon General' (Kluger, 1996).

The success of Marlboro established a model within the TTCs of deploying key brands backed with enormous promotional support as the motor of worldwide sales growth. The magnitude of the commitment involved is indicated by marketing expenditure in Asia. Malaysia has three tobacco companies among the top four marketeers, Philip Morris is ranked ninth in Hong Kong, an RJ Reynolds licensee is eighth in the Philippines, and the critical importance attached to gaining a foothold in the Chinese market is evident in Marlboro being followed by 555 State Express as the biggest advertisers in 1994 (Kaufman and Nichter, 2001). At the global level, Philip Morris spent US$813 million on overseas advertising in 1996 and was the world's ninth largest advertiser (Hammond, 1998b).

It is important to recognize that the term 'global brand' does not designate a globally uniform brand. Cigarettes of the same brand, but produced for differing markets, may vary significantly, for example, with respect to tar, nicotine and nitrosamine content (Gray *et al.*, 2000). In a similar vein, while core themes are evident in the marketing support given to key strategic brands on a global level, campaigns are not simply conducted in an undifferentiated fashion across all countries. In some cases, an element of national distinctiveness may arise from a perception of inappropriateness or cultural/commercial insensitivity. Although the Marlboro Country campaign has been successfully transplanted to many countries, in Hong Kong a prevailing disparaging assessment of cowboys was viewed by the Leo Burnett advertising agency as necessitating a cultural makeover, and the Marlboro Man consequently was depicted as driving a Jeep (Kluger, 1996). In Indonesia, a BAT study of smokers of international brands suggested that 'while the advertising concept can be standardized, deep-rooted local cultural differences seem to necessitate adaptation of the actual advertisements to the mores of smokers in Indonesia' (BAT Indonesia, undated).

In other circumstances, adaptation in international advertising campaigns may be the enforced result of local regulations. This may mean only marginal change to achieve regulatory compliance, as in China where a partial ban on cigarette advertising means that Marlboro billboards and television commercials do not include displays of cigarettes (Cunningham, 1996). A 1994 strategy document for Brown & Williamson's US international brands details how the image/advertising for CAPRI 'allows the female smoker to demonstrate her own individual sense of style. The international image campaign features contemporary, confident, and stylish women ... A campaign modification featuring couples is available for markets where advertising cannot be targeted at women' (Brown & Williamson, 1994).

More far-reaching innovation is increasingly required by the TTCs in finding ways to effectively communicate with potential consumers where regulations on cigarette advertising are gradually becoming more pervasive and restrictive. 'Brand stretching' or 'trademark diversification' has been prominent among the strategies employed to circumvent the restrictions that tobacco control legislation imposes. This denotes the development of new products or services to promote the image and awareness of a cigarette brand. In Thailand, for example, comprehensive controls on advertising have been partially undermined through such activities, with the promotion of Camel Trophy clothing being particularly prominent (Vateesatokit *et al.*, 2000). Malaysia's prohibition on television advertising of tobacco was easily side-stepped by the artificial construct of 'Salem High Country Holidays', via which fruitless attempts to book a vacation suggested that the operation existed solely to promote the Salem brand (Cunningham, 1996). BAT was a relative latecomer to this sphere of activity and its 1984 review of such operations by its competitors concluded that there are

> only a few activities which are serious trademark diversifications which are Marlboro clothing, Camel boots and Peter Stuyvesant holidays.... For the most part the other activities, whilst involving different products/services from cigarettes, are not commercially viable and seem designed to form a part of a brand's overall communication strategy and expenditure. (BAT Marketing Department, 1984)

Product placement has long been considered a supplement to conventional advertising and the film industry in Hollywood has provided a route by which TTCs have penetrated popular culture. The worldwide distribution of major Hollywood films means that such placement provides an opportunity for genuinely transnational brand promotion. Industry documents disclose an agreement by Sylvester Stallone to use Brown & Williamson products in five films for a fee of US$500,000, and payments by TTCs for the inclusion of their brands in films such as 'License to Kill' and 'Superman II' (Glantz *et al.*, 1996). A study of tobacco use in commercially successful movies notes a rising trend from background placements to potentially more powerful actor brand endorsements. This enhances the scope for the huge non-American audiences accounting for almost half of total revenues, to associate cigarettes with US culture (Sargent *et al.*, 2001).

TTCs have also been quick to exploit the possibilities presented by technological developments to circumvent regulation and increase brand awareness. Satellite television companies can broadcast across national borders on a regional basis and are thus capable of undermining national tobacco control measures. An important example is the Asian broadcaster Star TV, owned by the Philip Morris director Rupert Murdoch (Reynolds, 1999). In India the emergence of satellite television was used by the president of the India Tobacco Association to argue that a proposed ban on advertising was futile, since 'advertisements in this medium will offset the Indian ban' (*Tobacco News*, 1994b). BAT seems to have felt that their Indian associates were slow to realize the value of Star TV, with one executive suggesting that through 'its ability to show in pictures what previously was only available in words, I suspect that it will have a profound effect on aspirations, as it has elsewhere' (Greig, 1993).

The Internet is also viewed by TTCs as providing a powerful new medium, and rather covert strategies have been employed to exploit it. A leaked memorandum from BAT, for example, details Project Horeca by which the company was to invest £2.5 million (US$3.75 million) for a website aimed at young people around the world. The strategy was to provide a city guide to bars and clubs that would direct web-surfers to venues at which BAT cigarettes would be distributed (Simpson, 2001). A prototype site was launched in Poland and the scale of ambition for the project is indicated by the memorandum's stated minimum objective of '600,000 unique users by end of 2001' (Rogers, 2001a). Philip Morris has apparently used a similar strategy in Australia where a website offered free passes to a fashion event at which Alpine cigarettes were promoted (Harper, 2001).

5.3 Fast cars and cigarettes: tobacco company sponsorship of motor sports

Among the routes by which the tobacco industry has sought to maximize global awareness of its key product lines, particular importance is attached to the sponsorship of motor racing. Industry documents make clear the extent to which its core attributes have been identified as constituting an invaluable vehicle in the creation and promotion of desired brand images. Among the documents obtained from advertising agencies by the UK House of Commons Select Committee on Health, during its investigation into the conduct of the UK tobacco industry, was a 'qualitative debrief' regarding the brand Benson & Hedges and Formula One Sponsorship presented to the Gallaher company in 1997. The analysis concluded that:

- Formula One is one of the least contentious sports for association with cigarette sponsorship, indeed there is a natural fit between the two;
- The image is dynamic, macho and international, and consequently can potentially bring these image values to a brand; and
- Formula One is seen to be an appropriate fit for Benson & Hedges and can help to drive the more youthful and exciting elements of the brand imagery (Hastings and MacFadyen, 2000).

The value attached to Formula One within the global marketing strategies of the TTCs is further indicated by the reaction within BAT to the possibility of Brown & Williamson deciding not to renew its sponsorship of the Williams team in 1989. A report entitled 'Barclay Formula 1 Sponsorship' reviews the background to, and impact of, this involvement. The objectives attributed to this sponsorship include 'to add excitement and dynamism to the brand position', 'to enrich the brand's sophisticated, quality, upscale image' and to 'offer increasingly needed opportunities for additional means of communication'. In seeking to evaluate the results of this association, the report notes that:

It is impossible to 'prove' that a sponsorship is worth the money just as it is impossible to prove that an advertising campaign works. Each element of marketing expenditure should help to build a brand. BARCLAY has been supported with Formula 1 investment since 1983. Directly related to the F1 sponsorship or perhaps coincidentally, BARCLAY's market share has performed as follows:

	1983	1986	1988	5 months 1989
Finland	0.9	2.1	3.1	N/A*
Switzerland	3.4	4.6	4.8	4.8
Netherlands	1.2	2.6	3.4	4.1
Belgium	2.0	2.5	3.2	3.5

* 'Competition has stopped sharing sales. Barclay's volume is 2% ahead'. (Harding 1989)

This assessment of the advantages accruing from Formula One sponsorship also includes a discussion of how it performs by comparison with other high profile sporting events such as the Olympic Games, the football World Cup and major yachting races. Each of these potential alternatives is seen as having problems such as relative infrequency, regional confinement, uncertainty surrounding team qualification or an unwillingness to accept tobacco sponsorship. As stated in the report,

> Against this Formula One compares impressively:
>
> – The team always competes, so a sponsor knows his two cars will be on the grid at each Grand Prix.
> – TV coverage is massive around the world for each of the sixteen races.
> – There is a genuine association with the team, vital for image building. It is not a mere placard around an arena or track.
> – A sponsor will receive continuity of benefit as there will be sixteen races each and every year.
> – The races take place throughout the continents...
> – The media cover all matters Grand Prix throughout the year on television and radio, in newspapers, and the plethora of specialist and general magazines. (Hastings, 1989)

Of course, the TTCs have not confined their attention to Formula One. A wide range of motor sports sponsorships are designed to exploit regional variations in popularity among its various guises. BAT, for example, has used sponsorship of the Team Lucky Strike Suzuki motorcycle racing team to 'provide worldwide TV and Press exposure' and to 'provide int'l platform for in-Market exploitation and cross-promotional opportunities' (BATCo, 1993). The rationale for this involvement is strikingly similar to that employed in connection with Formula One, with perceived advantages of motorcycle research including 'substantial int'l TV audience', 'targeted to young adult males' and 'majority of races held in key Lucky Strike markets' (BATCo, 1993). In the US, leading tobacco companies have been heavily involved in the sponsorship of other motor sports including stock car, truck and Indy car racing (Spiegel, 2001).

The industry documents highlight several key issues in the utility to TTCs of involvement in motor racing. First, longstanding and widely held perceptions of the sport constitute the basis of its appeal to the companies. High-profile sponsorship is seen as enabling the association of particular cigarette brands with the more glamorous attributes of motor racing, with emphasis on the glamour, masculinity and dynamism that surrounds such events. A BATCo sponsorship tracking study for Spain emphasizes 'the good image fit between Lucky Strike target brand attributes and the image of GP motorcycle racing', particularly around image qualities such as masculinity, high quality, international, prestige and American values, while similar research in Indonesia indicates that 'the Lucky Strike sponsorship is driving the brand towards its target image' (BATCo, 1993).

Second, motor racing is perceived as a more effective way of enhancing brand perceptions than other sports of similar stature and public interest. This partly reflects intrinsic characteristics of the sport. The vehicles themselves allow sponsorship to be highly visible, thus enhancing the impact of the close association among the event, team and individual driver. In the UK, the Health Select Committee report notes that a key element in the 'synergy' of tobacco sponsorship with Formula One is the scope for the colouring and design of the cars to mirror that of the cigarette pack. This capacity is powerfully illustrated by the attempt of the British American Racing Team to race one car with Lucky Strike branding, and the other with State Express 555 (UK Health Select Committee, 2000). An additional core attribute of Formula One and motorcycle racing lies in the geographical spread of these competitions. Each functions as a worldwide contest, switching from country to country as the season progresses in a globetrotting whirl of television exposure and marketing opportunities.

Other facets of the relationship between the TTCs and motor racing are less to do with the intrinsic attributes of the sport and more with the comparative unwillingness of other sporting bodies to accept money from tobacco companies. Max Moseley, President of the FIA, Formula One's governing body, suggests in his appearance before the UK House of Commons Health Select Committee that the estimated £200–300 million (US$300–450 million) per annum received by the sport in tobacco sponsorship reflects a situation in which 'the tobacco people really have nowhere else to go' (Minutes of Evidence, 2000). As documented above, the involvement of TTCs in motor racing has generally attracted less opprobrium than other sporting pursuits that are more strongly associated with images of health, vitality and athleticism. However, it may be that certain events have added to the political sensitivities surrounding tobacco sponsorship of motor sports, particularly the so-called 'Ecclestone affair'. This concerned a £1 million (US$1.5 million) donation to the Labour Party by the Chairman of Formula One management, a donation that was widely perceived as contributing to the willingness of the Blair government to exempt motor sport from national and European measures to prohibit tobacco advertising and marketing (Bremner *et al.*, 1997; Wintour and Maguire, 2000).

An additional theme evident from the documents is the clear way in which sports sponsorship constitutes a means of circumventing restrictions upon advertising. As tobacco control measures have gradually imposed greater restrictions on more conventional forms of advertising, the TTCs have responded with innovative and expensive programmes to maintain their ability to effectively communicate with existing and potential smokers.

The scope afforded by motor sports sponsorship for such circumvention, and the enthusiasm with which the industry has seized this opportunity, are demonstrated by experiences in the US. Despite a longstanding ban on cigarette advertising on American television, exposure generated by motor sports sponsorships allows the TTCs to achieve an annual equivalent of over US$150 million in television advertising per year (Siegel, 2001). Industry documents emphasize that such evasion is an extremely cost-effective way of ensuring substantial television coverage. A 1989 review of BAT's involvement in Formula One notes that, from 16 March to 25 June 'the commercial value for BARCLAY added up to US$457,797 for 53 minutes and 26 seconds. BAT (B) is contributing only around US$200,000 to the sponsorship for the whole season' (Harding, 1989).

The involvement of the TTCs in motor racing is also indicative of the complex intermingling of global objectives with those at national or regional level. The broader global imperative pushing such sponsorship coexists with a number of geographically specific goals that differing elements in the overall marketing mix seek to target. The sophistication and precision with which such multiple targets are pursued is apparent in the attention given to the identities of individual racing drivers within sponsored teams. BAT's initial entry into Formula One with the Arrows Ford team apparently reflected an assessment of local advantage via the commodity value of a particular driver. Their Swiss subsidiary decided to sponsor Arrows 'because a Swiss driver, Marc Surer, was driving for Arrows Ford.... The Swiss driver provided the hook for local media coverage' (Harding, 1989). Having later switched to the Williams team in pursuit of the greater exposure associated with a more successful team, the review noted the potential gains of the rumoured arrival of Alain Prost as a driver for the following season: 'As he is French but lives in Switzerland he would be of even greater brand relevance than (Ricardo) Patrese' (Harding, 1989). It has recently been suggested that the Jordan racing team persisted with the services of the British former champion Damon Hill during a poor final season as a result of pressure from its major sponsor Benson & Hedges (Henry, 2001), the key brand of Gallaher, the market leader in the UK.

Above all, it is clear that the overriding purpose of motor sports sponsorship is the promotion by TTCs of strategically critical global brands. The UK House of Commons Health Select Committee Inquiry (2000) concludes that 'sponsorship is working exactly like advertising. The only significant difference between the two...is a disturbing one: the sales pitch in sponsorship is more hidden, enabling covert or "subliminal" messages that can get round the defences of their "wary" and media literate young targets.' The veracity of this assessment is readily confirmed by internal industry documents, as exemplified by an internal review of BAT sponsorships that states that 'Brand building is the key objective' (Hacking, 1993).

5.4 Creating the global smoker

While the attempts of TTCs to recruit non-smokers within low-income countries understandably receives the bulk of the attention and opprobrium of public health advocate, it is clear that the enormous opportunities for new cigarette sales are not confined to them. In particular, consumers of traditional forms of tobacco such as bidis, kreteks and chewing tobacco represent an enormous potential market. This is especially so in the context of increasing trade liberalization, a further means of promoting what is referred to as the 'global smoker' (Yach and Bettcher, 1999). One tobacco trade publication has described the continuing predominance of traditional products within key markets for expansion, such as India and Indonesia, as illustrating the limits of globalization, but it is made clear that these barriers are being aggressively targeted:

> For how long will these markets resist the attraction of global trends? In one or two generations, the sons and grandsons of today's Indians may not want to smoke bidis or chew pan masala. Cigarette manufacturers seem not to be asking if, but how fast these markets will change. Global brands are one way to accelerate this process. (Crescenti, 1999)

Industry documents demonstrate the determination with which the major companies are pursuing such a transformation in prevailing modes of tobacco consumption. For BAT the 1130 billion bidis consumed per annum in India represent 'a potentially lucrative source of business if they can be converted to cigarettes' (Burgess, 1994). BAT's principal Indian subsidiary, the India Tobacco Company (ITC), has sought to promote the western-style cigarette among the multiplicity of forms of tobacco use that make India such a diverse market. According to its Chief Executive, Kamal Ramnath, 'our primary aim is to expand the market for cigarette sales. We have the responsibility, being market leader, to do so' (Glass, 1997).

One key aspect of this campaign has been the attempt to equalize the taxation of bidis with that imposed on cigarettes. The manufacture of bidis employs large numbers of people involved in small-scale production; the governmental protection afforded by the political weight of the bidi lobby has been decried by the ITC. According to a former Chairman, it 'is because of the relative sympathy for the bidi industry that the Government has concentrated its hunger for revenue almost entirely on the cigarette industry' (Sapru, 1980). Attempting to rectify this perceived injustice, the ITC conducted 'dialogues with the Government at the Corporate and Industry level with a view to educating them about bidi smoking being a bigger problem than cigarette smoking, if smoking is to be considered a problem area at all' (Sapru, 1980).

The persistence of traditional forms of tobacco use is seen as a major obstacle to market growth for cigarette companies. There are indications that BAT has felt that the ITC has not always responded satisfactorily to the challenge presented by bidi smokers. Following a visit to the ITC in 1994, BAT executive David Aitken reported to Ramnath the opinion that the

> ITC need to better understand the biri [bidi] market in order to encourage uptrading to cigarettes. You must understand consumer reasons for smoking biri which may include image, the taste, feelings of naturalness and ease of putting out/relighting in addition to price factors.... ITC is undertaking considerable product development work but needs to create innovative products, especially to encourage uptrading from biri. (Aitken, 1994)

While much of the attention of the TTCs has focused on the taxation issue, BAT has argued that the 'ITC now needs to examine the psychology of bidi smokers to formulate the best approach for promotion and conversion' (Davis, 1994).

The international attributes conferred upon western-style white stick cigarettes have been deployed as a key resource for increasing market share. At governmental or societal level, TTCs have sought to present the rise of cigarette sales as an indicator of modernity and symbol of economic progress within low-income countries. A letter from the ITC to the Minister of Health similarly emphasized that 'any Government initiative must first encourage a conversion to cigarettes, which is the internationally accepted form of tobacco use' (Chugh, 1992). The promotion of cigarettes has been accompanied by veiled and highly dubious health claims, claims particularly pronounced with reference to international brands. To individual smokers, the transition to such brands has been presented primarily as an indicator of personal success and prestige. The strong association of 'global' or 'international' with 'high quality' has been used to imply a health benefit to the consumer that accompanies the switch from cheap local cigarettes to a quality brand (Whidden, 1990). In Indonesia, a BAT study of smokers of global brands found that one of the perceived advantages of white cigarettes was that 'it is less dangerous to health' (BAT Indonesia, undated).

The transnational nature of the TTCs themselves has also been critical in their expansion in many low- and middle-income countries. This is evident in their heavy reliance upon global imagery and appeal in promoting key brands, but this core element of the globalization of the tobacco industry should not be reduced to a simplistic 'Americanization'. There are clearly strong elements of this and the association with American lifestyles and values has been central to the global success of Marlboro in particular. BAT's assessment of Marlboro's growth in the Netherlands is that the brand 'capitalised on the move towards greater individualism and freedom which was a strong social trend in the 1970s' and exploited 'the popularity of Americana and associated imagery' (BAT, 1992). The success of Marlboro has consequently been treated as something of a paradigm for the expansion of cigarette brands into new markets and the strong association with American values of freedom, strength and prosperity is a recurrent theme in advertising and marketing cigarettes. The ITC's marketing plan for the launch of Lucky Strike on the Indian market similarly sought 'to build excitement and hype around the entry of the first USIB (US international brand) into India in an age of liberalisation and globalisation of the Indian economy', with the plan of focusing attention on the 'International/American pedigree and connection' of the brand (ITC, 1994).

A BAT study of consumption of international brands in Indonesia found that, for smokers of Lucky Strike, the 'link with America gives the smokers a feeling of pride' (BAT Indonesia, undated). The attempt to associate cigarettes with images of freedom and liberty have been particularly stark in Eastern Europe and the former Soviet Union. In the Czech Republic, for example, an advertisement for L&M cigarettes features a picture of the pack alongside the Statue of Liberty with a slogan, translating as 'This is what America tastes like! New Arrival!' (Cunningham, 1996). When R.J. Reynolds sought to attract young Japanese smokers via its now notorious Joe Camel campaign, Japan Tobacco responded by appropriating a key figure in American popular culture, introducing the Dean brand named after the all-American film icon James Dean (Frankel, 1996).

The assumed superiority of global brands, and particularly American products, is clear in the frequent preference for smuggled over locally produced versions of such brands. In Vietnam, Philip Morris insists that the local version of Marlboro is equivalent to those manufactured elsewhere, but many Vietnamese still prefer foreign produced cigarettes. Indeed, there is a certain cachet attached to the 'Made in the USA' label that appears on contraband products (Dreyfuss, 1999). BAT's study of Indonesian smokers of global brands partly attributes their preference to their association with greater prosperity and education, which 'makes them prefer (sic) brand with the Internationalism element (Even, they prefer the contraband, supposed to be "original from abroad")' (BAT Indonesia, undated).

The industry has, however, been far more sophisticated in its handling of issues of national identity than a crude Americanization hypothesis would suggest. The marketing of brands has been far more specifically targeted to specific population groups, and cultural ties and national images have been used in a diverse manner. The powerful imagery associated with America has been strategically critical to the expansion of global brands, but what is striking is the complexity and diversity of images and values that the TTCs have co-opted, along with the carefully nuanced manner with which they have been deployed. The distinction between USIBs and UKIBs, for example, designating brands originating from the US and the UK respectively, is viewed as commercially significant and, therefore, should be apparent to the consumer. This is evident for BATUKE (United Kingdom and Export) regarding a company proposal to voluntarily include health warnings on packages for markets currently without regulations requiring them. Their response emphasizes that the choice of wording 'should be appropriate to nationality of trademark owner. Consumers will likely seek to identify US trademarks from UK trademarks and will need re-assurance, therefore differences in English wording will be required' (Lovett, 1992). It was noted that the imposition of such a warning could 'prove to be a useful marketing tool in so much that it would endorse the internationalism/European origins of the product' (BATUKE, 1992). This contrasts with a concern expressed in Nigeria that the inclusion of health warnings would 'cause severe problems (perception of local manufacture) if the product bearing a Nigerian health warning is offered to consumers' (Mulligan, 1993).

America is not the sole repository of images, perceptions and values with which the TTCs wish to align themselves. In parts of Asia, Japan has been accorded a model status similar to that given to the US in Eastern Europe. Japan's economic success, in particular, makes the country a signifier of prestige, globalism and prosperity. In planning the development of the State Express 555 brand, one BAT executive emphasizes that:

> The key is for us to create a presence in Japan for this brand because of the growing importance Japan plays in the Asia Pacific region. As you know, the stature and health of a brand is critically influenced by who smokes it and if a brand is seen to be smoked by Japanese we create a strong synergy with the franchise growth in the region. This is because more and more Japanese travel the region and are seen as a group to set 'Western/International' standards. (Rembiszewski, 1994)

The extreme paucity of tobacco control regulations in Japan, in combination with its regional significance, has also allowed industry advocates to employ it as an exemplar in other parts of Asia. In India, for example, an article published by the Tobacco Institute of India offers Japan as a model for the relationship between government and the tobacco industry. It claims that the 'protection of the tobacco interests is one of the starkest examples of a national strategy the Government has pursued vigorously for decades – putting manufacturing might first, while often regarding consumers as a national resource' (*Tobacco News*, 1994a). A letter to the Indian Minister of Health from the Chief Executive of ITC again advocates learning from Japan as 'the foremost economic super-power of the Asia-Pacific region, and a model for all developing countries' and notes that 'a Japanese Government Report of '91 initiated by their Ministry of Finance, states categorically that no causal relationship has been established between smoking and health' (Chugh, 1992).

The degree of tobacco industry interest in exploiting or appealing to national identities extends to targeting individual brands at specific population groups. One example is the decision of Gallaher to sell Chinese cigarette brands in the UK. With the apparent aim of targeting Chinese communities in major urban centres, Gallaher signed an agreement with the Shanghai Tobacco Company to distribute 'Chungwa' cigarettes, China's leading premium brand. As reported by *The Times* newspaper (2000), 'smokers with fond memories of lighting up on Shanghai's Bund will be able to buy a packet of 20 Chunghwa at Gerrard Street in London's Chinatown'. Targeting of this kind has long been practiced in the US where mentholated cigarettes have been targeted disproportionately to African-Americans, and brands named 'Rio' and 'Dorado' have been aimed towards Hispanics (US Department of Health and Human Services, 1988).

Research assessing the potential market for State Express 555 among Chinese-Americans illustrates the nuanced way in which cultural identities and migrant aspirations are exploited to promote cigarettes (Asia Link Consulting Group, 1991). Conducted in San Francisco in Mandarin and Cantonese among respondents who have lived in the US for up to seven years, the BAT study highlights difficulties encountered by the community including the sense of isolation, language barriers and discrimination. It also notes that freedom is the primary motivation for migrating to the US. Given the success of State Express 555 in China, the 'premium, well established and quality image' of the product within the country of origin is seen as providing a significant business opportunity if properly reinforced. The study concludes that

Advertising to Chinese Americans in their language is pivotal:

- Impressed by advertising
- Personal relevance
- Emerging market

The advertising strategy was then further refined through micro marketing 'to maximize impact by strategically placing media within areas that possess a high concentration of Chinese-Americans', including the tactic of developing a presence in targeted Chinese areas of New York and Los Angeles during Chinese New Year celebrations.

5.5 A global contest: dominating discourse, resisting regulation

Notwithstanding the often proclaimed ferocity of competition among TTCs, the frequency with which the companies advance arguments in concert seems striking. Analysis of internal industry documents reveals examples of what might reasonably be termed conspiracies, with clear co-operation among the companies to advance their shared interests. The extent of such co-operation among ostensibly fierce rivals justifies the view that, for the TTCs, the primary competition is not among themselves, nor even with smaller national manufacturers, but with advocates of tobacco control. For companies such as Philip Morris, BAT and Japan Tobacco, their shared stake in resisting the further spread of comprehensive regulation is more central to continued industry success than the relative fortunes of key brands. It is clear that the companies view this as a genuinely globalized conflict, with the passage of effective tobacco control measures in any one country potentially 'raising the bar' in other parts of the world. As such, and given the magnitude of the economic stakes involved, the TTCs have engaged in wide-ranging efforts to control the manifold debates surrounding tobacco and health.

One of the most fundamental components of such efforts has been the attempt to manage the very terms on which such debates are conducted. The concern with language has been longstanding and, while it is most frequently associated with the protracted denials of a causal link between smoking and lung cancer, such concern both persists and covers many diverse aspects of how the TTCs operate and are perceived. At a superficial level, this is evident in the contemporary mantra of corporate responsibility propounded by tobacco companies. Brown & Williamson, for example, proclaim via its website that 'we are committed to demonstrating that Brown & Williamson is a responsible company in a controversial industry' (Brown & Williamson, 2001). In a similar vein, a series of advertisements by Gallaher carry the tagline 'An international tobacco company behaving responsibly', a motto that is also given prominence on the company's website (Gallaher, 2001). Listed among the highlights of 2000 in BAT's annual review is the assertion that 'we have made good underlying progress towards improving our acceptance as a responsible tobacco group' (BAT, 2001a) including £3.8 million (US$5.7 million) of funding to create an International Centre for Corporate Social Responsibility at the University of Nottingham (Tysome, 2000). Philip Morris is reported to have recently commissioned the London advertising agency Doner Cardwell Hawkins to develop a European campaign that will position the company as more socially responsible (Rogers, 2001b). Such apparently innocuous posturing has significant implications as the basis of concerted efforts by tobacco companies to remould the way they are perceived by the public and decisionmakers. This effort is geared towards recreating their widespread acceptance as acceptable businesses in society, often with a subtext of having rectified errors in previous behaviour. The 'new responsibility' represents an attempt both to legitimize their operations, and to add credence to the diverse arguments advanced to protect them.

A similar logic underpins current industry positions on addiction. Apparent admissions on the part of TTCs that nicotine is addictive have received much media attention (BBC News, 1999), but such admissions merely represent attempts to defuse the issue via its trivialization. Cigarettes are acknowledged as being addictive, but only according to some conceptions and in a banal habit forming sense presented as equivalent to many other routine activities. Hence, according to the Chairman of BAT Martin Broughton (1999),

> On the matter of addiction, there are several definitions in use: under some, smoking, as well as coffee drinking and also chocolate eating, is addictive. While stopping smoking can be difficult for some, we do not consider that there is anything in cigarette smoking that removes the ability of someone to quit, as evidenced by the millions who have.

On a similarly contingent basis, Japan Tobacco International accepts that smoking is addictive 'as the term addiction is used today', but insists that 'equating the use of cigarettes to hard drugs like heroin and cocaine, as many do, flies in the face of common sense' (Japan Tobacco International, 2001). Gallaher (2001) likewise acknowledges that 'in today's language, smoking is regarded as addictive', the term now being given 'such a wide interpretation that it encompasses a range of behaviours, including smoking'. These formulations are, at best, attempts to extricate the companies from the cul-de-sacs into which they had placed themselves by their earlier adherence to unconvincing denial. More fundamentally, such interpretations serve the critically important role of allowing cigarettes to be presented as a product like any other, albeit one with some risks attached, consequently deflecting attention away from their unique threat to global public health.

The TTCs have also sought to shape the discourse within which more policy specific issues are contested. Advertising represents a stark example of how tobacco companies have sought to switch the terms of debate from the public health focus of tobacco control advocates. The industry has long sought to portray the question of prohibitions on advertising as being one concerned with freedom of expression, maintaining that bans represent an infringement of commercial speech rights. Faced with proposed legislation banning advertising in the UK, BAT insisted that such bans 'limit fair competition amongst companies for smokers' custom, and limit the rights of companies – manufacturing and marketing legal products – to participate fully and fairly in their lawful business' (BAT, 2000c). Japan Tobacco similarly asserts that 'the freedom to advertise is fundamental to any consumer goods company's ability to compete for market share' (Japan Tobacco International, 2001). Tobacco companies have invested large sums in organizations prepared to advance such arguments, with Philip Morris and the International Advertising Association contributing almost US$1 million to the American Civil Liberties Union (Yach and Bettcher, 2000).

A powerful indicator of the success with which the TTCs have frequently managed to establish the parameters within which policy debates are conducted is smuggling. Despite mounting evidence of the extent of their complicity in smuggled cigarettes, estimated to account for around one-third of total exports worldwide (Joossens, 1998), TTCs have been able to convince policymakers that smuggling is a product of 'excessive' taxation of tobacco products. Japan Tobacco International insists that the 'root causes of contraband are high taxes, the unforeseen consequences of trade and regulatory controls, and inadequate law enforcement' (Japan Tobacco International, 2001), Brown & Williamson (2001) claim that 'High taxes and import restrictions or bans cause smuggling', while Gallaher (2001) asserts that the 'high taxes imposed in some countries create a situation where cigarettes and other tobacco products are increasingly being smuggled into these countries'. This claimed link is of immense significance to the tobacco companies since increased taxation is an effective means of reducing consumption (Ranson *et al.*, 2000; Sunley *et al.*, 2000).

The corollary of the claim that levels of contraband are shaped by tax differentials, and the primary value of smuggling as a means of exerting political leverage, is the assertion that national revenues would fall in countries that seek to improve public health via increased taxation of tobacco products. Industry arguments, backed by strong media pressure, have provided the basis for changes in taxation policy in countries such as Canada and Sweden. Progressive increases in Canadian cigarette taxes between 1979 and 1994 brought a dramatic fall in per capita cigarette consumption and significant increases in tax revenues. These public policy gains were reversed when an industry orchestrated campaign to induce and highlight awareness of a rise in contraband from the US led to a roll back in taxation in 1994 (Cunningham, 1996). The imposition of higher taxes in Sweden in 1996–97 again brought the dual benefit of generating further revenues and falling consumption, but limited evidence of a marginal rise in smuggled cigarettes brought a reversal in both tax policy and the associated gains (Joossens *et al.*, 2000).

Such cases add substance to the increasing evidence that there exists a virtuous circle in tobacco control policy, by which increased taxation of cigarettes brings gains to both public health and government finances (Jha and Chaloupka, 1999). However, this increased awareness does not yet seem to have curtailed the ability of the tobacco companies to prevent such policies from being adopted. The UK government recently reversed its previous commitment to successively increased taxation above the rate of inflation in the context of media interest in increased smuggling, with the move presented as a reinforcement of government attempts to control contraband and protect revenues (Bennett and Blackwell, 2001).

An increasingly significant area in which the tobacco industry seeks to structure debate, and one of particular interest in the context of globalization, is the attempt to present tobacco control as an issue for high-income countries. This claim is often articulated so as to present the WHO as in thrall to the interests of western donors and NGOs, and thus unresponsive to the needs of poorer countries. This has been particular evident as WHO has developed its Tobacco or Health agenda. An investigation into industry attempts to undermine tobacco control activities at WHO included attention to what the industry regarded as the 'Third World Issue' (Committee of Experts, 2000). A workshop at a tobacco company conference in 1980 emphasized that 'third world issues can't be "left for tomorrow to deal with" since they affect the very basis of raw material supply' (Worster, 1980). Subsequent industry efforts to inhibit enthusiasm for tobacco control among low-income countries combined dubious economic analysis with assertions of irrelevance. A contact programme drawn up by BAT for the World Health Assembly in 1984 planned:

> To stress that the 'Tobacco or Health' question, as presented, is hardly a developing country question, but has rather been imposed and forced upon them by the industrialised world, without due consideration of the highly detrimental economic and social consequences this will have for them. (BATCo, 1988)

The same themes are currently being advanced in the context of industry attempts to undermine global support for the proposed Framework Convention on Tobacco Control (FCTC). In a speech to the International Chamber of Commerce, Martin Broughton (2001) warned:

> The necessary establishment of international rules and regulations is becoming a vehicle for those whose real interest is what I think can properly be called the 'New Colonialism'. By this I mean the increasing rise of western-based governmental and non-governmental organisations who believe their way is the only way. Like religious zealots they see the opportunity to convert the world. 'New colonialists' seek to impose the values of the developed world on the developing countries ... [and] this trend is hindering the socioeconomic advancement of the developing world by seeking to undermine their comparative advantage.

Such a view presents the FCTC as a threat to national sovereignty for low-income countries, with BAT commandeering the language of subsidiarity in proposing an alternative approach of leaving 'national governments free to develop the most appropriate policies for the specific circumstances of their country' (BAT, 2000b).

Perhaps the most prominent contribution to public health facilitated by the disclosure of industry documents has been the revelation of the extent of industry attempts to manipulate science. Financing research projects of dubious validity, disseminating convenient findings, organizing concerted attacks and discrediting of research showing the harmful health effects of smoking, and hiring of sympathetic scientists to perpetuate controversy or act as witnesses are among the tactics that have been employed. These tactics have been designed to limit the impact of scientific discovery on profitability, particularly via inhibiting the generation of more effective regulation of tobacco products and by avoiding liability for tobacco-induced illnesses.

The utility of the documents in this respect was established by a series of seminal articles published in the *Journal of the American Medical Association* in July 1994 following the disclosure of documents of the Brown & Williamson company by an anonymous source. These authoritatively revealed both the longstanding disparity between internal awareness and public denials of the health impacts of smoking, and the extent to which tobacco industry involvement in the scientific process was determined by commercial imperatives. The documents demonstrate that some of the industry's internally conducted research concluded that environmental tobacco smoke (ETS), for example, was harmful to health despite public denials that such hazards had been proven (Barnes *et al.*, 1995). Similarly, they illustrate that employees of Brown & Williamson and BAT undertook research on the pharmacology of nicotine, talking routinely and explicitly about addiction in direct contrast to the public positions of the companies (Slade *et al.*, 1995). Attention is further drawn to the role of lawyers in determining the content and conduct of industry-funded research, with scientific merit being of little significance in the selection of external projects to be funded. Instead, results were used to detract from the health dangers of tobacco (Bero *et al.*, 1995). Attorneys also appear to have routinely reviewed and revised scientific research prior to their publication (Hanauer *et al.*, 1995).

The great volume of documents released in 1998 as a result of the Minnesota Settlement Agreement has produced several studies disclosing the active co-operation of tobacco companies to shape the scientific process. Such collusion has often focused on health impacts of ETS as 'the most important single issue facing the industry' (Imperial Tobacco, 1987). This is well-illustrated by the industry's response to the largest European epidemiological study on lung cancer and ETS undertaken by the International Agency for Research on Cancer (IARC) (Committee of Experts, 2000; Ong and Glantz, 2000). In response, Philip Morris initiated a cross-industry grouping that involved BAT, R.J. Reynolds, the British company Imperial and Germany's Reemtsma.

PM initiated and chairs an industry-wide task force to manage both the IARC monitoring and scientific intelligence gathering process and the development of a global communications/government relations plan to address [the] impact of the study. (Winokur, 1994)

The amount of resources allocated by the industry for this task is indicative of its perceived critical importance. While the IARC study itself cost between US$1.5–3 million over a ten-year period, Philip Morris alone budgeted US$2 million in 1994 alone for its response, and proposed another US$4 million to fund studies to discredit IARC's work (Ong and Glantz, 2000). BAT appears to have taken the lead in instigating an international programme of press briefings that served to defuse the impact of the study even prior to its publication, with the industry promoting the view that the study demonstrated no increase in the risk of lung cancer for non-smokers.

The global nature of such collaboration is notable. 'Project Whitecoat' was a programme initiated by Philip Morris by which independent scientists were recruited by the law firm Covington and Burling (Dyer, 1998). The project encouraged the participation of other companies:

Philip Morris presented to the UK industry their global strategy on environmental tobacco smoke. In every major international area (USA, Europe, Australia, Far East, South America, Central America and Spain) they are proposing, in key countries, to set up a team of scientists organized by one national coordinating scientist and American lawyers, to review scientific literature or carry out work on ETS to keep the controversy alive. They are spending vast sums of money to do so. (cited in Drope and Chapman, 2001)

In Asia co-operation took the form of the 'Asia ETS Consultant Project' aimed at recruiting scientists that could be trained up to testify on behalf of the industry in legislative, regulatory or litigation contexts. This programme placed a strategic premium on identifying local scientists throughout the region:

This objective was based on the fact that there were essentially no local scientists with a background in ETS issues and that experience elsewhere has shown that it is essential to have credible, local scientists prepared to speak out when ETS becomes an issue, which often occurs on short notice. (Rupp and Billings, 1990)

In other situations, the ability to 'parachute' scientists in from prestigious western research institutions has been seen as advantageous in handling local responses to the ETS issue. A conference on occupational hazards and air pollution, for example, was identified by Philip Morris as an opportunity for generating valuable media attention. In discussing media strategy for the event, Charles Lister of Philip Morris suggested invitations to industry scientists George Leslie and Roger Perry, both of Imperial College, London:

> I suggest interviews, joint or separate, with Leslie and Perry. They play off each other well, and between them cover both health and IAQ [indoor air quality] questions. I can imagine stories to the effect that two eminent British scientists are in Cairo to attend an international conference regarding air pollution, and kindly expressed their views regarding those issues. (Lister 1993)

This again also highlights sensitivities to the manipulation of national identities and global imagery, and illustrates the scope for adaptation to local circumstances within the parameters of a global strategy. Industry documents show that the TTCs have long been attuned to the opportunities for 'spin', casting artificially constructed doubt and controversy to deflect the attention of legislators and the public from broad scientific consensus on the adverse health impacts of tobacco.

5.6   Conclusions: the global public health response

Much of the above is clearly discouraging from the perspective of global public health, but it is important to recognize that the tobacco industry does not have monopoly control over the means by which cognitive globalization is occurring. While acknowledging the huge advantages conferred by the industry's considerable resources, the global power of the TTCs has been far from unchallenged. Indeed, the cognitive dimension of globalization brings certain opportunities for promoting and protecting global public health.

Although there remains a clear disjunction between the global character of the tobacco industry, and the essentially national nature of tobacco control measures, the successful development of comprehensive measures in countries as diverse as Canada, Thailand, Australia and South Africa offer positive prospects elsewhere. This is clear from the nature of the industry's reaction to proposed legislation in these countries, namely an overriding concern with a potential 'domino effect'. This is explicitly recognized in BAT's response to a proposed advertising ban across the EEC:

> Both BAT and the industry (through CECCM) have set themselves the objective of ensuring that individual European governments retain the right to determine their own advertising restrictions (bearing in mind that an EEC advertising ban will inevitably extend to Eastern Europe and have a detrimental effect elsewhere in the world). We believe it is essential for BAT to take urgent and positive action to try and persuade the Commission not to issue any new directive for the following reasons:
>
> – Protection of existing freedoms in the whole of Europe (east and west) is essential for the progress of international brands.

- Advertising in eastern Europe is already threatened; an EEC ban would have an immediate impact.
- The global 'domino effect' – an EEC ad ban will speed up the introduction of bans elsewhere. (BAT 1991)

Similarly, a key area paper distributed to general managers of BAT's European countries regarding indirect taxation of tobacco products warned that 'International cooperation and cross-fertilization between Governments is increasing with regard to indirect taxation, as a result of which both taxation levels and structures are being shaped by external influences' (BAT, 1995).

One specific manifestation of this phenomenon of policy spread is the adoption of measures in a forerunner state to serve as models elsewhere. This has been described as '"leap-frogging" – i.e. a process in which various countries have pioneered effective measures, which have then been progressively adopted elsewhere' (Framework Convention Alliance, 2001). This is typified by the spread of advertising bans following their adoption in Scandinavia in the 1970s, while Thailand's Tobacco Product Control Act clearly used earlier Canada legislation as its prototype (Vateesatokit, 1997). Historical or cultural links between particular countries may also encourage policy spread. For example, the India Tobacco Company notes the tendency of legislators to use the laws of the UK, Australia, New Zealand and Canada 'as the basis for fresh legislation in India' (Misra, 1988).

The tobacco industry's fear of policy spread is also apparent in its response to global collaboration among tobacco control activists. In a guide produced by INFOTAB (International Tobacco Information Centre), an international industry coordination group, entitled 'World Action! A Guide for Dealing with Anti-Tobacco Pressure Groups', the organization seeks to provide national manufacturers associations with an early warning alert system and a pro-forma action plan to counter any increases in NGO activities. The document demonstrates the extent of industry concern at the crossborder mobility of ideas and personnel associated with activist groups. Among the danger signals to recognize, the guide suggests, are the 'presence of *activist group(s)* and *key individuals*', as well as

The starting up of an antis' *coalition*, and its development, e.g. Canada, New Zealand, Scandinavia. Especially dangerous is the calling of a press conference by the coalition. Coalition recognition characteristics:

- professional/career activists with an anti-tobacco background
- medical/health groups as core
- ethical flash-point. Once a certain number of medical groups join, then others will follow
- mobility, e.g. use of the New Zealand Toxic Substances Board report in Norway by National Council on Smoking and Health Chairman

(Bjartveit), a red-hot activist. Also used in US Congressional Hearings (original italics) (INFOTAB 1987; 601005627)

In planning to counter NGO activities, the guide urges national associations to be aware of the 'global implications of one country's regulations cascading through others – effects on marketing freedoms, intellectual property and volume' (INFOTAB 1987; 601005631).

The advances in communication technologies that the tobacco industry has sought to exploit are also available for the use of tobacco control advocates, albeit on a less lavishly funded basis. The Internet, in particular, allows for more rapid and effective interchange of information and expertise across national borders. A particularly prominent example of such co-operation in the field of tobacco control is GlobaLink, an electronic network managed by the International Union Against Cancer. At its outset, an American spokesman for the Tobacco Institute suggested that 'this expanding, sophisticated infrastructure is probably one of the greatest challenges our industry has confronted' (Yach and Bettcher, 2000). It now links over 1200 tobacco control programmes worldwide, providing advocates with access to databases, news bulletins, archives and electronic conferences.

The expanding global impact of tobacco litigation in the US also constitutes an important facet of cognitive globalization. In part, this connotes the adoption by other countries of litigation as a direct means of achieving public policy goals. Countries such as Guatemala, Venezuela, Bolivia and Nicaragua are pursuing third party reimbursement cases in Washington DC (Daynard *et al.*, 2000). While the ultimate value of the spread of such litigation remains contentious (Jacobson and Warner, 1999), the success of the Minnesota case in releasing internal industry documents has provided an unprecedented opportunity for advancing global public health.

Finally, and perhaps most importantly, the negotiations facilitated by WHO for a Framework Convention on Tobacco Control (FCTC) hold the potential to transform the scope for the globalization of tobacco control. While media attention has understandably focused on the frustrations of the public health community at positions taken by key governments, the FCTC process itself has already had diverse and interesting impacts. Changes and innovations have been introduced allowing the more active participation of civil society, notably the presence of many NGOs at negotiating sessions as observers and the holding of extensive hearings reminiscent of the US Senate. This has encouraged more intensive and organized co-operation among NGOs from different parts of the world in support of the FCTC, evident in the formation of the Framework Convention Alliance (Collin *et al.*, 2002). The breadth of active participation among WHO member states is requiring some engagement with tobacco control issues in countries where regulation has remained minimal. In Zimbabwe, for example, the establishment of a National Task Force to develop a framework for tobacco control is at least partially attributable to this process (Woelk *et al.*, 2000). While negotiating an effective and comprehensive FCTC is clearly a daunting task, to say nothing of the enormity of securing compliance, the FCTC retains the potential to become a major global public health movement (WHO, 1999a). The prospect for such innovation in the strengthening of global health governance is much needed to tip the balance in the global challenge to stem the tobacco epidemic.

Notes

1. The defendants were Phillip Morris Inc., R.J. Reynolds Tobacco Company, Brown & Williamson Tobacco Corporation, B.A.T. Industries plc, Lorillard Tobacco Company, American Tobacco Company, Liggett Group Inc., Council for Tobacco Research and the Tobacco Institute.
2. Many of the references for this chapter indicate tobacco industry documents that have been released to the public as a result of litigation in the United States. Each page has had a unique identifying number allocated to it by the courts known as the Bates Number.

6
The Impact of Globalization on Nutrition Patterns: a Case-Study of the Marshall Islands[1]

Roy Smith

6.1 Introduction

There is a growing body of literature considering linkages between nutrition patterns and global processes of social change. Popkin (1994) highlights this with specific reference to low-income countries. This work is later developed by Drewnowski and Popkin (1997) who argue that patterns, such as the changing relationship between income and fat consumption, are best understood within the context of broader global trends. Pricing and the availability of vegetable oils, fats and sweeteners, together with increased urbanization, are seen as relevant factors in the nutrition transition. Lang (1999) takes these arguments further by drawing attention to the manner in which marketing techniques and governance of food systems increasingly operate at the global level.

The work referred to above has the common theme of linking selected patterns and processes of globalization with an increased risk of poor diets and diet-related morbidity and mortality. A central aspect of definitions of globalization used is that the individual is increasingly affected by interrelationships far beyond their immediate locality. However, Drewnowski and Popkin (1997, pp. 40–1) acknowledge the dual importance of individual actions as health determinants and more structurally-defined factors, writing that '[t]he globalization of the human diet appears to be the combined outcome of innate sensory preferences coupled with the greater availability of cheap fats in the global economy and rapid social change in the lower-income world'. Similarly, Lang (1999, p. 341) argues that '[W]ider considerations such as European and US trade enlargement are likely to raise rather than reduce global tensions in the food sector. The challenge of how to balance seemingly contrary policy imperatives (health, environment, consumer aspiration and commerce) and how to bridge tensions within the food system (land, industry, retailers, catering and domestic life) is formidable.' Such a challenge can only be successfully addressed with an understanding of the nature of the relationship between individuals' diet, health and broader socioeconomic factors.

This chapter analyses the global economic and social factors that contribute to unhealthy dietary patterns through a case-study of the Marshall Islands. This small island state is an especially useful case-study because, as a relatively small and geographically isolated community, the factors influencing dietary habits and other aspects of social change are more readily identifiable than among larger, continent-based communities. There have also been changes in dietary patterns, notably the increased consumption of imported processed foods, that can be related to the particular influence the US has had on the Marshall Islands and other former UN Trust Territories of the Pacific Islands. Although some aspects of this influence are unique to these islands, they also represent a microcosm of significant core–periphery relations within the evolving global political economy.

The chapter focuses on Marshallese perceptions of the nature of good health, in relation to diet, and how economic, cultural and environmental factors are seen to influence their health. These perceptions suggest that orthodox explanations of dietary patterns, and the prevalence of diet-related health status, offer an insufficient understanding of the health dynamics at play. Rather, it supports explanations that show how global economic and social changes have created conditions at a micro-level that shape dietary patterns. Better understanding of these trends in the Marshall Islands provides lessons for the impact of globalization on health determinants in other countries.

6.2 The Marshall Islands

The Marshall Islands comprise 29 coral atolls and five small islands scattered across 750,000 square miles of the northeast Pacific Ocean. The population is currently around 65,000 with the majority located in the two main urban areas of Majuro and Ebeye. These islands have been influenced to a greater and lesser degree by the colonial forces of Spain, Germany, Japan and, under the United Nations' Trusteeship system, the US. Although formally an independent republic, the Marshall Islands' political status has been one of Free Association with the US since 1986. This relationship has resulted in a significant flow of direct grants and federal programmes from the US to the Marshall Islands. Over US$1 billion have been allocated between 1986–2001, amounting to 75 per cent of the Marshallese budget. The effect of this has been a strong economic dependence on the US.

As with many societies and economies in transition to modernization, and integration with an increasingly global political economy, the Marshall Islands has experienced a mixture of costs and benefits associated with this process. This is a problematic field of study in that assessing such changes is inherently subjective, with respondents having differing attitudes and value systems. Also external observers may have an over-romanticized perception of life in remote Pacific islands. For the islanders themselves, many appear to welcome the opportunity to migrate to the growing urban conurbations with a view to participating as fully as possible in the evolving patterns of consumerism.

In a small Pacific country like the Marshall Islands, where traditional lifestyles were relatively healthy, it might be expected that good population health would be readily achievable. However, some aspects of modernization and globalization have contributed to economic, social and cultural factors that support unhealthy lifestyles. These include external influences such as a global economy, that is often presented in terms of liberal free trading, yet actually has significant structurally defined imbalances embedded within it. Underlying and reinforcing such imbalances is the sometimes overlooked, yet fundamental, aspect of cash-based systems of exchange. Once a society has embraced such a system, it is at the less-than-tender mercy of market forces. As implied above, these forces favour certain groups over others.

Despite investment in health care during the period of Trusteeship (1947–86), life expectancy at birth has increased slowly in the Marshall Islands to current figures of around 60 years for men and 63 for women. Infant mortality rates have fallen significantly since before World War II, although still relatively high at 63 infant deaths per 1000 live births in 1988, falling to 27 per 1000 for 1996 (McMurray and Smith, 2001, pp. 102–7).[2] It is significant that continuing low life expectancy can be attributed to the offsetting of gains in child survival and reductions in infectious disease burdens by a high prevalence of noncommunicable diseases. The key causes of morbidity and mortality are now cancers, cardiovascular disease and diabetes. Thirty per cent of the population over 15 years of age, and over half the population over 50 years of age, has diabetes. Surgery in the Marshall Islands mainly consists of diabetic foot amputations.[3]

A 1951 nutrition survey of Marshallese diets found that all age groups were deficient in calories, minerals and vitamins. Notably, this followed disruption to local food production caused by World War II, and marked the beginning of dependency on imported foodstuffs. A similar survey in 1991 showed a marked contrast between older and younger Marshallese. The younger age groups continued to show evidence of under-nourishment and stunted growth. However, older respondents had increasing incidences of obesity with high rates of diabetes, hypertension, cardiac and osteopathic problems, and premature death. This strongly suggests that lifestyle factors are playing a significant role in how health is determined as Marshallese become older. As implied above, however, lifestyle choices are often structurally influenced by the options available to individuals.

To help understand the above trends in health status, and their relationship to nutrition patterns in particular, this chapter draws on qualitative research carried out in the Marshall Islands. Overall, the data provides important insights into Marshallese perceptions of health dynamics, and points to structural factors contributing to the high prevalence of diet-related illness in the Marshall Islands. Some of these factors affect health and health perceptions in other peripheral societies, although people living in other societies may respond in different ways. The Marshallese data also provide a useful illustration of the way perceptions of health of some groups may be inconsistent with some of the assumptions that form the basis of the Western model of health care.

Two main techniques were used to study Marshallese perceptions of health. One was a series of in-depth interviews that were carried out in Majuro with English speakers. The respondents were many and varied, including health professionals, high school teachers and other educators, former Peace Corps volunteers, journalists, clerics, politicians, administrators, youth workers and various Marshallese who spoke English. The second method was a series of focus group discussions carried out in Majuro by Marshallese interviewers with other Marshallese, using the Marshallese language medium.

A male and a female Marshallese interviewer were employed to carry out focus group discussions with small groups of three or four men or women in three age groups: adolescent high school students, 30–40 years, and 50 years and over. The free-ranging focus group interviews covered various perceptions of health and ideal body weight; the main health problems in the Marshall Islands; the link between diet and health; the causes of poor diets; non-dietary causes of poor health; maternal and child health; and stress and social causes of ill health. To avoid hindering the flow of discussion with potentially sensitive questions, and because the study focus was early-onset noncommunicable diseases (NCDs), the interviewers did not raise the topic of sexually transmitted diseases. However, focus group respondents spontaneously mentioned STDs and sexual behaviour from time to time as a health concern.

6.3 Existing understanding of unhealthy diets in the Marshall Islands

Explanations for particular health patterns can often be found in the characteristics, attitudes and experiences of populations. However, observers sometimes overemphasize the importance of some of the most obvious characteristics of populations, and may overlook more important explanations. This is evident in research on the Marshall Islands and other Pacific islands where it is common to attribute the high prevalence of lifestyle-related illness primarily to specific characteristics and experiences, and to underestimate the impact on health of global structural forces. Interviews with Marshallese respondents indicate that, in general, specific characteristics and experiences were less important as the causes of poor health than uneven economic and social development. Some characterizations even appeared to be incorrect. The following sections first examine four particular characteristics and experiences that are often overemphasized, followed by discussion of six factors that are more important causes of unhealthy lifestyles in the Marshall Islands.

Overemphasized factor 1: *Pacific people have different perceptions of health and beauty and they think ample body proportions are more desirable than a healthy weight*

Large body size is valued in many societies, and obesity has long been a symbol of high social status and prosperity throughout much of the Pacific, including the Marshall Islands. It does not automatically follow, however, that obesity is regarded as healthy or beautiful, or that this belief is a cause of the high prevalence of obesity in the Pacific region and an obstacle to the maintenance of a healthy bodyweight. Each group of Majuro respondents were shown sketches of men and women of different weights, equating roughly to ideal weight, 10 per cent overweight, 20 per cent overweight, 30 per cent overweight and 10 per cent underweight. In the sketches facial characteristics and stance were depicted as uniformly as possible so that the only major difference between respondents was their bodyweight. They were asked to state which of the figures was the healthiest (*ejmourtata*) and which was the most attractive (*emontata*).

Generally respondents selected the figures with approximately ideal weight as both the healthiest and the most attractive, although there were some small differences of opinion among the men whether the figures that were about 10 per cent overweight might be more 'cuddly'. Only one woman in the 50+ group selected an obese figure as the healthiest.

Respondent one: Well, some may say the man in the first picture is healthy because he is the fattest and others will say he is unhealthy because he is too fat.

Respondent two: Not the man in the first picture. He is too fat. And the man in the fifth drawing is not healthy because he is too thin.

Respondent one: I still say the man in the first picture is the healthiest because he is big and fat.

Respondent two: I say the man in the third picture.

Respondent one: The first woman looks big and healthy. And the second drawing is even better. Look at the size of her thighs! But the woman in the first drawing is the healthiest. The woman in the fifth drawing looks healthy too. But then we hear people saying people who are neither fat nor thin are healthy, so maybe the women in the third and fourth drawings are healthiest.

Respondent three: They are always telling us that we should lose weight, so, yes, maybe. They look like they might weigh the proper weight level – what is it, 125 pounds?

Interviewer: And what is not an ideal weight?

Respondent three: 140 pounds and over.

Respondent one: That fat woman will have health problems because she is too fat.

Respondent two: Those two slim ones will not be burdened by their weight.

These comments suggest that the respondents generally had a sound perception of health, and that most view excess weight as neither healthy nor attractive, especially among the younger age groups. It seems, however, that this is a change from views held in the past, as indicated by the views of an older group of women in the 30–40 age group:

Interviewer: I'm sure you remember from your childhood older people saying that someone who was healthy and prosperous was one who packed a few extra pounds. What are your thoughts on this?

Respondent: Well, we have now discovered that this is not the case. We were misinformed. Nowadays, through education, we know that an overweight person may not be all that healthy. The same may apply to someone who is very thin.

Thus, although obesity is still associated with high status, especially in older people, it is not necessarily perceived as healthy or attractive. As a symbol of status, rather than a factor defining status, obesity alone would not assist those who lack the rank and wealth to achieve high status.

In a subsequent interview, a Marshallese woman commented that Marshallese and other Pacific men often tell their wives that they look better when heavier. She stated that this was not because they themselves liked fat women, but to ensure that their wives would not appear attractive to other men. A Tongan woman expressed a similar view. This suggests that overweight and obesity are actually perceived as unattractive. Perceptions of ideal weight and health among the Majuro respondents do not appear to be greatly at variance with medically defined levels of ideal weight, and should not constitute an obstacle to the maintenance of a healthy bodyweight.

Overemphasized factor 2: *Pacific people do not know enough about major health problems and their causes; they need more health education, especially education in good nutrition*

Each group was then asked to name the main health problems in the Marshall Islands. The four adult groups – men and women aged 30–40 and

50 + – all mentioned diabetes as the leading health problem in the Marshall Islands. The four adult groups also mentioned cancer and thyroid problems. The adolescent girls and women aged 30–40 added kidney disease, heart disease, high blood pressure and complications of pregnancy to their lists of leading health problems. The adolescent boys mentioned diabetes but focused on the lifestyle habits of smoking, drinking, chewing tobacco, marijuana and cocaine. This is a clear reflection of their own experiences and current concerns, as well as their currently low risk of developing the most prevalent noncommunicable diseases.

A striking feature of the focus group discussions is that not one of the respondents mentioned an infectious disease as a leading health concern. Every condition discussed was noncommunicable and, with the exception of pregnancy complications, strongly associated with lifestyle risk factors. This reflects widespread community awareness and concern about the high prevalence of early-onset NCDs in the Marshall Islands. The high levels of such diseases, which are more likely to be an underlying rather than an immediate cause of death, is attracting more attention than some of the leading immediate causes of deaths, such as pneumonia and sepsis.

Respondents were then asked their opinions about the causes of the health problems they had mentioned. All six groups of respondents ranked poor diet, and particularly the consumption of imported food, as a leading cause of ill health in the Marshall Islands. The adolescent boys and men aged 30–40 mentioned smoking and drinking and a poor diet of imported food as the major causes of such diseases. The adolescent girls also mentioned pollution and stress. Most were eager to discuss dietary problems in detail, and had very strong views on the subject, as indicated in these comments by male and female respondents in the 30–40 year age groups:

> *First man (30–40):* I firmly believe that the major cause of diabetes is the type of food people consume today, especially in the urban centres such as Majuro, Ebeye, and to some extent, Jaluit. And the type I am referring to here is *ribelle* (foreign) food … the food we import.

> *Second man (30–40):* As a young man in the late 50s, I do not remember a single case of diabetes here in Majuro. Back then people's diets consisted largely of local foodstuffs as there was only about one store on the whole island and a very limited shipping service. In comparison, today, with development of all parts of our society and our money-based economy, we consume mainly imported food items. And I have heard that 90 per cent of the patients in the hospital are diabetic. Don't you think there is a connection?[4]

> *First woman (30–40):* Diseases are caused by our consuming food that is not good for us, which may result in such a disease as diabetes. Most important for a person's health is the right kind of food. Fatty foods,

salty foods and sweet foods harm us. Fat, salt and sugar are the things that cause health problems in the Marshallese.

Second woman (30–40): We consume too much sugar, especially those of us residing on the central islands who have no access to such traditional foods as coconut sap. Our diets consist mostly of rice, as we have very little access to our traditional foods. This is the main reason there is such a high number of diabetic cases today. We have chosen *ribelle* food over our own. We all know there is [a] greater number of people with diabetes today than long ago.

Male respondent (30–40): Food items such as corned beef, turkey tails, rice, sugar and salt are harmful to us. But the way we prepare our food is also a factor. We all know that frying is a favourite Marshallese way of preparing meats of all types. Now we're being advised that excessive use of fat in the preparation of food, as well as consumption of fatty foods can lead to such health problems as diabetes.

Woman (50+): Our traditional food helped us in other ways too. For example, brushing one's teeth was taken care of by chewing pandanus. The fibres cleaned our teeth. So people long ago had healthy teeth. Nowadays, you see people who are still young who are wearing dentures. I personally believe people long ago were healthier because of the food they ate.

The women aged 50+ attributed the generally better health of *ribelles* to their different diet, and drew some interesting comparisons:

Respondent one: Ribelles are really healthy because they eat what is good for them, they eat the right kind of food. They have clean surroundings and eat at the proper times and eat the proper amount. Whereas, Marshallese eat anything, anytime, and in large, large amounts.

Respondent two: Yes, all we eat is store bought. So we feel sick and unhealthy. It seems *ribelles* do not eat a lot of meat. They eat mostly greens.

Respondent three: That's true. Things like cabbage. But we do not know how long most of the *ribelle* food we buy to eat has been stored so we cannot select the best quality. Besides they are not our foods so we do not know how to prepare them properly.

Respondent two: Yes, ribelles know what goes well with what to get the most out of what they eat.

It is clear from the comments of these Majuro residents that lack of dietary knowledge is not the main cause of high levels of noncommunicable diseases. Respondents in all age groups were well aware that their general level of health is poor and had a good understanding of the relationship between nutrition and health. However, a concerted nutrition education programme has brought little improvement in Marshallese diets. As discussed below, socioeconomic factors play a much greater part in influencing population health.

Overemphasized factor 3: *Marshallese attribute most of their health problems to nuclear testing and do not realize they are related to lifestyle*

Another common explanation for the persistence of poor lifestyle habits in the Marshall Islands is apathy as a consequence of the extreme insults to community health inflicted by the US nuclear testing programme. It is argued that this leads them to underestimate the contribution of their own lifestyle habits to NCDs such as diabetes and cardiovascular disease.

The effect of radiation on the health of those Marshallese who were exposed to nuclear testing is widely recognized, both officially and by the Marshallese population as a whole. There is little doubt that radiation has been a significant cause of NCD in the Marshall Islands, including various cancers and thyroid conditions. However, the focus group interviews indicate that generally respondents have a good appreciation of which diseases are primarily a consequence of radiation, and which are more closely associated with lifestyle. The four adult groups all mention radiation as a cause of disease. In each case, they also cite the contribution of diet and lifestyle factors to noncommunicable diseases, even those generally attributed to radiation. An example of this balanced view is the following exchange in one of the male focus groups:

> *First male respondent (30–40)*: Besides the type of food we consume, I think some of the health problems we are experiencing today are a direct result of the nuclear tests conducted by the United States on Bikini and Enewetak. I know for a fact that a number of the major illnesses afflicting Marshallese today were nonexistent before the testing programme. The Nuclear Claims Tribunal has drawn up a list of various cancers that fall under its compensation plan.

> *Second male respondent (30–40)*: May I interrupt? While I agree that the nuclear tests may be a major factor in regards to health problems, I also would like to point out that the Division of Public Health informs us that such diseases as cancer are a result of cigarette smoking. So again, this is yet another indication that imported items play a significant part in people's health and wellbeing.

The simultaneous appreciation of the real and grave impact of radiation on Marshallese health, and also of the contribution of lifestyle factors to

NCD, is particularly well-illustrated by the remarks of one woman in the 50+ age group:

> The *ribelles* contaminated our islands with their experiments, causing us to get cancer, thyroid, and other diseases. Another cause of the high rates of cancer and thyroid is our inability to buy the kinds of food that will not cause us to get these illnesses.

In fact, no hard evidence currently exists to demonstrate that diet can protect from, or reduce the risk of, radiation-related cancers, leukaemia and thyroid conditions. This respondent apparently assumes an association exists because she has been made very aware that a healthy diet greatly reduces the risk of developing diabetes.

One popular belief that constitutes a possible over-attribution of problems to radiation is that nuclear testing has caused coconuts and fruit to become stunted and produce less fruit:

> *Woman (30–40):* Diseases [such] as cancer and thyroid are a result of these *ribelles* bombing our islands. The whole of the Marshalls is contaminated. And so you have people with thyroid, cancer; the direct result of poison from the bombs. Our flora are also affected, their growth is stunted; they too are contaminated. These are the problems.

Although there is scientific evidence to indicate that coconut, pandanus and breadfruit in the northern atolls contain radioactive substances, including Caesium 137 (Maragos, 1994: 70), aging of trees is a more likely explanation for declining yields.

Overemphasized factor 4: *Pacific people have trouble staying thin because of 'the thrifty gene'*

A fourth commonly offered explanation for poor Marshallese health relates to obesity. It is often argued that the Pacific physiology is distinctive, in that it is better adapted to a feast–famine cycle because of 'the thrifty gene'. This particular genotype confers a survival advantage where marked fluctuations in food availability exist, such as in the traditional Pacific lifestyle. People with this gene, including most Pacific and Australian Aboriginal populations, are very efficient at storing nutrients in the form of fat reserves, which can be burned up when food intake decreases. Such people are thus more able to store nutrients during 'feast' periods and more able to survive 'famine' situations. This genotype becomes a disadvantage, however, when food availability tends to be more uniform, such as in a modern urban environment where there are no famines. Those who are efficient at storing nutrients store too much when exposed to a continual diet of high fat, high sugar, low fibre food. If periods of food shortage are not encountered, fat reserves simply accumulate and people become obese (Maragos, 1994: 55).

Although there is strong evidence that the thrifty genotype contributes to Pacific obesity, the main cause of obesity is eating patterns, not a specific genotype. The thrifty gene may make it more difficult for people to lose weight through restricting food intake, but this does not prevent mainte- nance of a healthy bodyweight if eating habits are adjusted. Substantial numbers of Marshallese are not overweight or obese, even though they share the same genetic heritage. It is also worth noting that improved metabolic efficiency can be developed by almost anyone who severely restricts calorie intake. Weight lost in periods of semi-fasting is quickly regained when normal eating patterns are resumed, and it then becomes increasingly difficult to shed weight solely by dieting.

It is thus evident that the above commonly cited explanations of the high prevalence of NCDs in the Marshall Islands, as attributable to particu- lar perceptions of health or genotypes, are not supported by the data of this research. It is argued that, in order to identify more fundamental causes, it is necessary to look at the broader political and economic setting, notably the impact of globalization on the Marshall Islands and Marshallese response to these forces.

6.4   Global factors contributing to unhealthy diets in the Marshall Islands

Urbanization, employment and diet

Some of the most important causes of poor health habits identified by respondents relate to low incomes, a separation from traditional lifestyles and displacement from the means of subsistence production. In the Marshall Islands, urbanization has occurred without industrialization or substantial generation of employment. This has resulted in displacement from the means of subsistence production, coupled with a widespread inability to pur- chase nutritious food. Majuro and Ebeye contain 67 per cent of the total population, but many residents rely on a share of the royalties from the American military presence or the income of employed relatives. According to government statistics, in 1996 13 per cent of working age Marshallese were unemployed and 79 per cent of the unemployed were aged 15–29 years. Many of the 25 per cent reporting as self-employed may have been under-employed. The employment situation has worsened since the mid 1990s as a result of population growth and public sector cutbacks. At the time of writing, unemployment on Majuro was approximately 20 per cent.

In relation to diets, low-income groups are able to purchase only the cheapest food items. In the Marshall Islands this means white bread or pol- ished white rice as the staple food, supplemented with low-grade fatty cuts of meat such as frozen turkey tails and spare ribs. The most popular and cheapest convenience foods, such as chips (french fries), tend to be high in fat and low in nutritional value. Sweet foods and carbonated drinks are also available at low cost, whereas fruit and fruit juices, are relatively expensive. Perishables that are not produced locally, including meat, fresh fruit and vegetables, must be air-freighted to the Marshall Islands. Thus, retailers stock only small quantities to minimize spoilage and sell at a higher price to cover additional costs. Finally, because of the absence of a local farming industry, low-income families are generally unable to supplement these staples with healthy quantities of fresh fruit, vegetables and proteins (for a further discussion of the economics of the nutrition transition see Thaman, 1988).

Man (30–40): We not only consume only imported food stuffs, but are restricted to consuming only what we can afford. But now the information we are getting from the Ministries of Health and Social Welfare is that these types of food are not only non-nutritious, but can also lead to health problems if consumed in large amounts. But what are we to do if these are all we can afford?

First woman (30–40): Marshallese eat mainly rice, bread, chicken, beef, tuna and other canned meats. These are the cause of our health problems today, the reason for so many types of illnesses. A typical Marshallese meal would consist of boiled rice, meat and perhaps a pot of tea. Not many people would think or bother to add other food items such as cabbage or papaya. Sometimes it may be that they have not acquired a taste for cabbage and sometimes they cannot afford the cost of cabbage.

Second woman (30–40): Residing on Majuro prevents our having access to suitable traditional food. Sometimes traditional foods may be found in the stores, but we do not possess the financial means to purchase them. While prices in stores are continually going up, the minimum wage has remained constant. This is another problem. Now our government has been reducing the work force.

First woman (50+): A place like Rita has no pandanus, very few coconuts and breadfruit trees. There are not enough places to grow these plants. People on the outer islands have healthy skins from the food they eat. But people on Majuro have scabies and other skin diseases and many are also night blind.

Second woman (50+): So when we get sick we go to see the doctor and are told we need to eat cabbages and other greens to help us control our diabetes, but the problem is, where do we get the money to buy the cabbage and other greens? And where can one find space to plant on Majuro? If we were living on our own island, then we could plant a garden. People on the outer islands do not have as high a rate of diabetes and cancer as do we on Majuro, and they live longer.

Third woman (50+): Even if you do have a place to grow these foods in Majuro, people steal what you plant. That's how much people realize our own foods are better for us than *ribelle* food.

Urban poverty and displacement from the means of subsistence production is a worldwide phenomenon among groups marginalized by globalization (Bettcher *et al.*, 2000). In almost every country, populations that migrate to urban areas, whether to find wage jobs, share in the benefits of modernization or flee political unrest, are at risk of poor nutrition, substandard housing and unsanitary surroundings. Their capacity to live a healthy lifestyle is largely determined by their ability to earn cash income. When such incomes are low, there is little choice but to purchase the cheapest food which, as described above, tends to be the least nutritious. This is especially true in a country such as the Marshall Islands where almost all food in urban areas is imported.

Globalization, mental health and unhealthy lifestyles

The stresses associated with urbanization itself are a leading factor in unhealthy lifestyles. In the poorer parts of urban Majuro and in Ebeye, settlements are dense and *over crowded*.

Adolescent girl: Houses are small and crowded together in Rita and Ebeye with many occupants and it's hard to get peace and quiet to study or sleep. Often you can't get to sleep because other people in the house or the neighbours are making a noise. Sometimes if we have male relatives around and we need to go to the toilet we will endure the discomfort until they are away from the bathroom entrance because it would be immodest to let them see you enter. I think we harm our health by suppressing a natural urge.

Not only do such circumstances generate stress, but they offer no outlets for it. Attempts to externalize feelings, such as by expressing anger, can lead to social conflict, whereas repression and internalization of anger increases personal stress. People who have no ready outlet for stress-induced feelings may overeat, smoke or drink alcohol as a way of expressing frustration. Domestic violence, including both spouse and child abuse, is also a symptom of stress.

One of the main causes of stress in urban Marshallese is economic insecurity. Budgetary assistance from the US contributes more than 80 per cent of the Marshall Island's gross domestic product (UNDP, 1994, p. 46). Many people living in urban areas fear they could be severely affected if assistance is reduced substantially and there are further cutbacks in employment and deterioration in economic conditions. Any cutbacks that occur would impact more severely on those living in urban areas than people on the outer islands who still have the opportunity to practice a subsistence lifestyle.

> *Man (50+):* Living on the centres is difficult because you need money to survive. I cannot get what I need to survive if I do not have a job. And since I do not own the property I live on, if I want a coconut, I need money to buy it. Whereas, if I lived on my island I would not need money to survive. If you have only one person employed in say, a household of 15 people, about the only thing that person can buy is rice. A small amount can feed all members of the household. In order to buy better food, that person would have to earn more money.

> *Second woman (50+):* Money, or the lack of it, is the root of all our social problems today. If people had enough money they would not have to worry so much and would be able to buy all the healthy and nutritious food their bodies require. This is why we have so many murder cases, especially among young, unemployed men. They may want something but do not have the means to obtain it so they get drunk and kill and rob someone, so they can buy whatever it is they want.

Although there have been relatively few murders in the Marshall Islands in recent times, another symptom of extreme stress that has become more prevalent is suicide. In 1994 there were ten suicides, and between 1 October 1995 and 30 September 1996 there were nine suicides, all committed by males (MOHE, 1996, p. 58; MOHE, 1997, p. 25A). They represent around seven per cent of all male deaths in each year. Most theories about the high rates of suicide in Micronesia and elsewhere in the Pacific attribute them to stress and conflict caused by changes in the traditional social structure, or lack of opportunity for self-fulfilment (Rubenstein, 1992).

The promotion of imported foods by nutritionists and retailers

In the Pacific region, nutrition tends to be viewed mainly as a health issue, with little attention paid to the importance of the availability of food, and patterns of production and distribution (Schoeffel, 1992, p. 241). However, retailing patterns and food availability are crucial determinants of health risk in the Marshall Islands where few people produce their own food. The interaction between consumer demand and retailer supply is complex and circular. On the one hand, consumers must have knowledge of available products before they exert demand for them. On the other hand, retailers must have effective demand in order to supply a product. Small Pacific countries offer limited markets to retailers, discouraging them from stocking a wide variety of goods or introducing new goods that may not be assured of a market.

Consequently, food retailers in the Marshall Islands tend to stock a range of foods based on the American diet rather than on a traditional Pacific diet. This includes processed and packaged foods, white flour products, convenience foods, carbonated soft drinks and alcohol. Initially these foods were introduced to satisfy the demands of American military personnel resident in the Marshall Islands. When the Marshallese acquired a taste for such foods, retailers supplied this larger market. Food aid also contributed to dietary change in the Marshall Islands. American-oriented nutrition programmes did not recognize the value of local foods such as green coconut milk and encouraged the consumption of imported foods such as dairy produce and orange juice (Schoeffel, 1992, p. 240).

Given that established retail patterns are unlikely to change without corresponding changes in consumer preferences and effective demand, it is likely to be difficult to introduce improvements to diets on the Marshall Islands. In August 1997 one of the principal Majuro retailers began selling traditional foods at lunchtime on Fridays and Saturdays. He reported that this was popular and sales were good. However, the purchase price was beyond the means of poorer Majuro residents, and as sales were only intermittent, this activity is unlikely to be a spearhead for dietary change. Interestingly, respondents recognize the role of retailers and the profit motive in shaping, not only their food preferences, but also the prevalence of drinking and smoking in the community.

> *Woman (30–40)*: Aren't cigarettes and alcohol two of the largest money-making imports today? Some people know that drinking and smoking are bad and yet persist in indulging. Maybe the parliament should pass a law banning the import of alcoholic beverages and cigarettes. It would not be bad and such a law would force people who use the two substances to stop.

The question of whether Majuro residents would willingly reduce consumption of fat, sugar and salt in order to change to a healthier, traditional diet if given the option is an interesting one. Although focus group discussions indicate that all age groups know this would be healthier, preferences for imported foods are now well established. Majuro residents clearly enjoy high-fat, sweet and salty food. Several informants mention that Majuro residents send store-bought food such as rice and canned beef to relatives on outer islands because they believe they will be enjoyed. Children are fed on sweets and snack foods because some mothers believe that they must be good, because they are expensive and children like them. The standard fare served at restaurants and take-away outlets is very rich – typically comprising large servings of fried fish or poultry, coleslaw, white rice or potato salad drenched in mayonnaise. Locally baked white bread is noticeably sweetened and liberal amounts of sugar are normally added to hot beverages. These preferences are reflected in the consumption patterns of imported food-stuffs, rising in volume to account for 30 per cent of all Marshall Islands imports by the mid-1990s. With preferences for such foods so well established, and people unaccustomed to eating a wide range of foods, it is difficult to encourage healthier diets. Attempts to do so would need to be gradual so that taste preferences can adjust over time. The words of one respondent indicate how resistant Majuro citizens can be to the overzealous promotion of a traditional diet:

> *Man (30–40)*: I personally feel it is too late. We cannot very well advise people to revert back to a subsistence type economy. I mean, who would be willing to give up rice, corned beef, tea and sugar and go back to existing on breadfruit, fish and coconut water?

Local perceptions of community and mortality

In the Marshall Islands, as elsewhere in the Pacific, community bonds are strong and community values are central to culture and daily life. Whereas modern Western society tends to promote individualism, in Pacific societies it is socially unacceptable to place individual needs ahead of community values. These values may also be influencing health patterns. In the Marshall Islands, the majority of people are practising Christians who believe the soul is immortal and that those who have lived a good life will one day be reunited with their families in Heaven. This, along with the belief that God cares for his flock, may encourage passivity as regards self-care (Simmons and Voyle, 1996, p. 103).

Moreover, the Christian churches teach that those who go to Heaven will be free of the troubles and suffering of life on earth. In a communally oriented society these basic Christian beliefs tend to support the attitude that it is more important to nurture community bonds which will endure beyond death, than to adopt an individual view. Death itself may be perceived not so much as an end but as a transition to a better place. This greatly reduces the fear of death, and makes it less important to take individual action to be healthy in order to avoid an early death, but very important to observe community values and maintain a secure place in one's community. In this context it would appear foolish to risk alienating the community by behaving in a socially unacceptable way, such as by refusing food and drink or moving in an undignified fashion, simply to avoid an early death.

Assessment of personal health risks

In both low- and high-income countries, many people can have a poor understanding of the causes of disease, and do not appreciate the direct links between unhealthy behaviour and noncommunicable disease. This does not seem to be the case in Majuro, where respondents generally demonstrated a sound appreciation of the causes of noncommunicable disease. This indicates that health education programmes are effective and are reaching their target audience. However, in Majuro, as elsewhere, knowledge alone is insufficient to prevent people engaging in unhealthy lifestyles.

One reason is that, even if people understand the adverse health consequences of some behaviour, they may assess their personal risk as low. The underestimation of personal risk from NCDs such as cancer, heart disease and diabetes is largely because they take many years to develop and do not affect everyone who is at risk. Hence the connection between unhealthy behaviour and disease is less direct. It has been estimated that smoking reduces life expectancy at age 20 by seven minutes per cigarette. However, most indulge for many years before they develop lung cancer, for example, and some people never do. Consumption of five alcoholic drinks in one day is estimated to reduce life expectancy at age 20 years by an hour, and each drink in excess of five reduces it by a further 20 minutes (Manning *et al.*, 1991, pp. 62, 86). Yet the consumption of large amounts of alcohol, like a diet too rich in fats and sugars, does not have an instantaneous impact on health. This does not mean that the statistics are incorrect; but rather that they refer to average risk and not to a definite outcome. Very different behaviour might be forthcoming if the connection between lifestyle choices and disease were more immediate (e.g. if all smokers developed lung cancer one week after they smoked their first packet of cigarettes). Since this is not so, it can be very difficult to use risk arguments to persuade people to abandon activities that give them immediate pleasure. This is especially true in a culture such as the Marshall Islands where social interaction is particularly important, of which eating, drinking and smoking are an integral part.

Thus, Marshallese who are well informed about the health risks attached to certain behaviours do not necessarily modify their behaviour. Like people everywhere in the world, if it is difficult to give up certain behaviours and there is not strong support from family and friends to do so, they can persuade themselves that they will be one of the lucky ones who escape the longer-term consequences. Alternatively, people pushed into retreatist behaviour by adverse economic and social circumstances may continue with deleterious behaviours even when well aware of the risks. In such cases, deliberate self-destructive behaviour, such as drinking to excess, may occur regardless of individual perceptions of risk. Such behaviour is evident not only in the Marshall Islands, but in many other peripheral countries and marginalized groups throughout the world.

6.5   Conclusions

Between the 1950s and 1990s the Marshallese have experienced two significant health trends that are indisputably connected. First, there has been a marked shift in lifestyle patterns with regard to diet and exercise. Second, the levels and causes of morbidity and mortality are closely correlated with these lifestyle patterns. Higher intakes of fat, sugar and salt have directly resulted in early onset diabetes and related illness and disease. Similar connections can be made with a range of cancers and cardiovascular illnesses. It may appear that if the main reasons for the current patterns of morbidity and mortality are lifestyle-related, then it would simply require adequate health education programmes to alter lifestyle preferences.

The discussions with Marshallese residents described here, however, make it clear that the major forces shaping health-related behaviour in the Marshall Islands are strongly related to structural changes linked to globalization, rather than any unique characteristics of the population itself. It is not a lack of health education that is at the heart of this problem. Quite the contrary, the Marshallese appear to be particularly *well informed* with regard to the determining factors relevant to their health. Rather, a key impact of globalization on the Marshall Islands has been displacement from the means of subsistence production, coupled with a widespread inability to purchase nutritious food. This experience of urbanization without industrialization or substantial employment generation is a key underlying factor affecting the high prevalence of NCDs in Majuro and Ebeye. In turn, the health of residents of the outer islands is affected by the underdevelopment of the country as a whole. Moreover, economic change has led to increased stress and social disruption.

It would be wrong to suggest that the Marshallese, or any other community, can completely abdicate any personal responsibility for their own lifestyle choices. The peripheral location of the Marshall Islands as regards the US is a major determinant of the lifestyle options available to Marshallese, although cultural factors, for example, influence choices from the available lifestyle options. Yet, this research shows that there are significant structural factors at play that fundamentally impact on the available options and subsequent choices made by the Marshallese. Somewhat ironically the Marshallese, as with other economically peripheral or marginalized communities, are experiencing the double-edged impact of globalization. On the one hand, they are being drawn into an increasingly global political economy. On the other hand, they are structurally disadvantaged in their relationship with other states and the global business community. It remains to be seen how this relationship will develop. Current indications are that the Marshallese are at least aware of the disadvantages they face. What is more difficult to predict is whether they will be enabled to achieve a balance between enjoying the benefits of globalization, while at the same time acting on choices that reject or resist its unhealthier consequences.

Notes

1. This chapter is based on empirical research carried out in the Republic of the Marshall Islands by the author and Christine McMurray and published in *Diseases of Globalization*, Earthscan, 2001.

2. There has been a degree of variation in the annual number of infant deaths, which suggests it is affected by chance fluctuation. Therefore it is difficult to discern a long-term trend.
3. The Pacific Diabetes Resource Center, www.pdtrc.org.
4. This is a substantial overstatement but a figure that was frequently mentioned by Marshallese not employed in the health sector.

7
Globalization and the Challenge of Health for All: a View from sub-Saharan Africa

David Sanders and Mickey Chopra

7.1 Introduction

In 1978 the concept of Primary Health Care (PHC), codified in the Declaration of Alma-Ata, was explicitly outlined as the key strategy for achieving Health for All (HFA) by the year 2000 (WHO and UNICEF, 1978). This strategy aimed to provide for basic health care needs, as well as address the underlying social, economic and political causes of poor health. Certain principles underpinned the strategy, namely universal access and coverage on the basis of need; comprehensive care with an emphasis on disease prevention and health promotion; community and individual involvement and self-reliance; intersectoral action for health; and appropriate technology and cost-effectiveness in relation to available resources. These principles were consonant with the 'basic needs approach' of development discourse during the 1970s, which emphasized social investment, integrated development and collective self-reliance (Walt and Vaughan, 1981).

Since the heady days of the late 1970s, however, progress towards HFA has been problematic. Foremost has been the resurgence of conservative economic and political forces during the 1980s, resulting in a marked paradigm shift in international health towards economic rationalism and technocratic solutions. This perspective permeated development discourse as a whole and profoundly influenced approaches to health policy and international aid. Importantly, this shift has been closely aligned with the promotion of neoliberal forms of globalization which, contrary to the goals of HFA, have led to a decline in health status of certain population groups as reflected in basic health indicators.

This chapter will analyse progress towards achieving HFA since 1978, measured in terms of health outcomes, in relation to economic, social, political and health systems factors linked to globalization. Using Zimbabwe as a case-study, rapid changes in the above factors and their impacts on key determinants of health, are illustrated. The chapter concludes that only a return to more equity-based policies focused on comprehensive PHC can meet the challenge of HFA in a globalizing world, and reverse the recent downward trends in health status in sub-Saharan Africa (SSA).

7.2   Mixed progress in world health[1]

Over the past 50 years, considerable gains in health status have been achieved worldwide. Life expectancy at birth has increased from 46 years in the 1950s to about 65 years in 1995 (WHO, 1998b), and total deaths among young children have been reduced to about 12.5 million instead of a projected 17.5 million (UNICEF, 1996). Substantial control of many communicable diseases, notably poliomyelitis, diphtheria, measles, onchocerciasis and dracunculiasis, has been achieved through immunization and other control measures (Tarimo and Webster, 1994), and cardiovascular diseases have decreased among males in high-income countries, in part due to declining smoking rates (WHO, 1998c).

Despite these gains, there have also been setbacks. Although in aggregate terms child mortality and life expectancy have improved in all regions of the world, disaggregation of data reveals that the gap in mortality rates between high- and low-income countries has widened significantly for certain age groups. The relative probability of dying for under-fives in low-income countries compared to western industrialized and eastern European countries, for example, increased from a ratio of 3.4 in 1950 to 8.8 in 1990 (Legge, 1993). In a number of countries in SSA, infant mortality rates (IMR) actually increased in the 1980s as a result of drought, war and civil unrest, HIV/AIDS and economic recession (Commonwealth Secretariat, 1989).

The past two decades have also seen an alarming resurgence and spread of 'old' communicable diseases, once thought to be well under control, including cholera (see Lee and Dodgson in this volume), tuberculosis, malaria, yellow fever, trypanosomiasis and dengue. In addition, 'new' epidemics, notably HIV/AIDS, threaten the last century's health gains in many countries. Many low-income countries are also experiencing an 'epidemiological transition', with cardiovascular diseases, cancers, diabetes, other chronic conditions and violent trauma replacing communicable diseases in some social groups, in other groups coexisting together as a 'double burden'. Some term this process an 'epidemiological polarization', with disadvantaged population groups experiencing high burdens from both communicable and noncommunicable disease (Frenk *et al.*, 1989).

7.3   A global shift in paradigm: from comprehensive to selective primary health care

The factors influencing this mixed progress in health development are varied and complex. It is argued in this chapter that one key factor has

been changes in the development discourse, manifested in health development as a shift in paradigm from comprehensive to selective PHC. Since
1978 there have been important successes in implementing PHC. During
the 1980s, significant progress in improving coverage for some population
groups with essential elements of health care was achieved. The most
impressive achievements have been in child health through the vigorous
promotion of selected 'child survival' interventions such as growth monitoring, oral rehydration therapy (ORT), breastfeeding and immunization.
Of these, immunization has led to the most dramatic improvements, with
global coverage of children under one year increasing from 20 per cent in
1980 to 80 per cent by 1990 (WHO, 1992b). Such energetic efforts to promote child survival are described as 'selective PHC', delivered by vertical
programmes using technocratic solutions, in contrast to 'comprehensive
PHC' focused on the broader determinants of health (that is income
inequality, environment, community development) and appropriate technology (Rifkin and Walt, 1988). Beginning as a difference of perspective
between WHO and UNICEF, the debate soon widened with selective PHC
finding favour with the growing conservative political ideology of the
1980s that advocated a minimalist role for states, reduced public expenditure and hence targeted health interventions. The result was the enthusias-
tic promotion of such interventions to the detriment of broader approaches
involving communities and other sectors (Chopra *et al.*, 1998).

Selective PHC was given further credence by the World Bank's *World
Development Report 1993: Investing in Health* (World Bank, 1993), which recognized the importance of health to economic development. Based on calculations of 'disability adjusted life years' (DALYs) and 'burden of disease'
by country and region, the report defined a core package of health care
provision from what was deemed the most cost-effective interventions.
This concept of the 'core package' then became a mechanism for limiting
the health services provided by the state, with nonstate (not-for-profit and
for-profit) organizations playing a greater role. Importantly, this process of
priority setting fitted neatly with the Bank's wider approach to macroeconomic reform, which included reducing the size of the state through privatization of public services.

An assessment of selective PHC suggests short term but unsustainable
health gains. There is growing evidence that immunization coverage is
stagnating worldwide, declining since 1995 in SSA and since 1996 in Asia
(Simms *et al.*, 2001). Infant mortality rates are again rising in many SSA
countries (Gadomski *et al.*, 1990; UNICEF, 1996). Various project evaluations have raised questions about the sustainability of mass immunization
campaigns (Hall and Cutts, 1997), the effectiveness of health-facility-based
growth monitoring (Chopra and Sanders, 1997), and the appropriateness
of ORT when promoted as sachets or packets, without corresponding
emphasis on nutrition, clean water and sanitation (Werner and Sanders,
1998). A systematic review has pointed out the lack of evidence for the

effectiveness of directly observed therapy for tuberculosis (DOTS) in the absence of well-functioning health services and community engagement (Volmink and Garner, 1997). Other studies have found that, only where health interventions are embedded within a comprehensive health care approach, including attention to social equity, health systems and human capacity development, can real and sustainable improvements in health status be seen (Fitzroy *et al.*, 1990; Halstead *et al.*, 1985).

7.4   The global context: globalization and health in sub-Saharan Africa[2]

The links between health and income distribution are well documented – the greater the inequalities in wealth within and across countries, the greater the differences in health status (Wilkinson, 1992). Much evidence suggests that the chasm between rich and poor counties has widened to record extremes. The United Nations Development Programme (UNDP) reported in 1993 that the one-fifth of the world's population living in the wealthiest countries had 83 per cent of the world's gross domestic product (GDP), while the poorest 20 per cent had less than 1.5 per cent (UNDP, 1993). By 1997 this gap had widened to 86 per cent and 1 per cent respectively (UNDP, 1999). Inequality in incomes has also increased within a diverse range of countries, as measured by the Gini co-efficient,[3] including China, Sweden, US, UK and countries of eastern Europe and Commonwealth of Independent States (CIS) (UNDP, 1999). Finally, inequalities of wealth are reflected across the world's population as a whole. The net worth of the world's richest 200 people doubled between 1994 and 1998, and the assets of the world's richest three people are more than the combined gross national product of the 600 million people of all the least developed countries (UNDP, 1999).

In a situation where one-quarter of the world's population (more than 1.3 billion people) live in absolute poverty, with an income of less than one dollar per day (UNDP, 1997), access to the most basic needs for good health is inevitably undermined. This reduced ability to meet basic needs within certain population groups is reflected in the following data. Between 1965 and 1992, daily per capita calorie supply increased from 72 per cent to 82 per cent of estimated daily requirements in low-income countries overall, but declined in SSA from 75 per cent to 67 per cent. The effects of inadequate dietary intake are further worsened by inadequate water and sanitation facilities, with nearly one billion people lacking access to safe water supply, and 2.4 billion people having no access to proper sanitation. In addition, in low-income countries, more than one billion people live without adequate shelter or in unacceptable housing (UNDP, 2001). Compounding these household level influences are environmental hazards, such as global warming and the dumping of hazardous products, which are impacting disproportionately on certain population groups

(WHO, 1998b). Furthermore, in low-income countries in general, and SSA in particular, economic insecurity and political instability are resulting in unprecedented humanitarian emergencies in size and complexity.

The causes of the above are the result of a complex history of uneven economic and social development extending over centuries, but progressing more rapidly over the past hundred or so years. What is notable, however, is the acceleration of economic globalization over the last three decades. The oil crises from the early 1970s and worldwide recession ended the postwar boom, the latter precipitated by stringent financial policies in high-income countries (particularly the US and UK) including tighter credit, higher interest rates and reduced public spending. The resulting economic slowdown impacted on low-income countries through a reduced demand for primary commodities and cuts in foreign aid (Raghavan, 1996). These developments, together with deteriorating terms of trade, led to a reversal in the flow of capital. Low-income countries, notably oil-importing countries, became net exporters of capital and burdened by huge foreign debts.

Among the most important policies introduced during this period, and intended to integrate low-income countries more closely into a globalizing world economy, have been Structural Adjustment Programmes (SAPs). Promoted by the International Monetary Fund (IMF) and World Bank, SAPs have resulted in significant macroeconomic policy changes including public sector restructuring and reduced social service provision. Any analysis of the impact of globalization on health necessitates an examination of SAPs and, it is argued here, their significant social disbenefits on certain population groups.

Zimbabwe is a useful case-study to link economic globalization, of which SAPs are a part, and health impacts. Zimbabwe experienced impressive health gains during the first decade following independence in 1980. Infant mortality (under one year of age) declined from pre-independence levels of 120 to 150 per thousand live-births, to 61 by 1990. Child mortality (one to four years) declined from 40 per thousand in 1980 to 22 in 1990 (UNICEF, 1994). These gains are attributable to the energetic implementation of new policies based on PHC. In particular, maternal and child health (MCH) services were vigorously promoted, with a massive expansion in PHC facilities and certain programmes such as the Expanded Programme on Immunization, ORT and community-based health care (Sanders and Davies, 1988).

Between independence in 1980 and 1991, the performance of the national economy fluctuated considerably. The early post-independence boom lasted two to three years and ended in 1982. The government briefly introduced, and quickly abandoned, IMF austerity measures. The economy remained protectionist, with expansion hampered by stringent controls on foreign currencies. Government debt grew. By the early 1990s, wage gains had been eroded by inflation and real wages remained at 1980 levels. Discussions with the World Bank and IMF began in the late 1980s and in

1990, although the economy was not in crisis, the government introduced the Economic Structural Adjustment Programme (ESAP). Its implementation began in earnest in March 1991 after a meeting in Paris with aid agencies and the World Bank. ESAP contained the standard features of IMF/World Bank reform strategies: a reduction of the budget deficit through cuts in public spending and rationalization of public sector employment; trade liberalization including price decontrol; deregulation of foreign trade, investment and production; phased removal of subsidies; devaluation of the local currency; and cost recovery in public services.

In an early review of the impact of SAPs on MCH, Costello *et al.* (1994) draw attention to the fact that earlier studies were largely based on retrospective and cross-sectional data, and to the need for more rigorous longitudinal studies. Being a relative latecomer to structural adjustment, Zimbabwe offered the opportunity for such research. The research subsequently carried out (led by one of the authors) sought to measure changes in health services and status during the implementation of the SAP. Through the monitoring of selected indicators of household economic performance, and utilization and functioning of health services, time series comparisons between equivalent seasons in successive years were made. The research extended from 1993 to 1998, and was conducted in one urban (Chitungwiza) and one rural area (Murehwa district).

Chitungwiza is a large conurbation situated about 30 km south of the capital city Harare. Few economic opportunities exist in Chitungwiza and many of the employed people commute to Harare on a daily basis. The official population of Chitungwiza is 274,912 (1992), of which 39 per cent are below 15 years of age and only two per cent are 60 years and older. Chitungwiza has one government hospital and four municipal clinics. Murehwa district is mainly comprised of communal farming areas and a small commercial farming area (Chitowa). The population is 152,505 (1992), 48 per cent are below 15 years of age and seven per cent are 60 years and older. Murehwa district has two hospitals and 12 rural health centres (RHCs), of which five are owned by the government, six by the rural district council and one by a Catholic mission. Data were collected through household surveys conducted annually in May–June between 1993–96 and 1998, and from hospitals and clinics in the two study areas.

Early in the implementation of the SAP a severe drought between 1991–92 led to reduced agricultural output, water and electricity shortage and reduced consumer demand. The resultant 12 per cent fall in real GDP per capita persisted for the next five years. The combination of devaluation of the local currency and removal of consumer subsidies on maize meal and bread, prescribed by the SAP, resulted in a dramatic rise in food prices and the consumer price index. The government estimated that price rises were 25 per cent higher for low-income families. A national poverty survey completed in 1996 showed that 64 per cent of Zimbabweans lived in poverty and about one-third (35.5 per cent) earned below the income needed to ensure adequate food intake (CSO, 1998). Further devaluations ensued: in June 1997, the Zimbabwe dollar stood at ZWD12 to US$1, by November 1998 it had fallen to ZWD40 to US$1. Rising inflation, estimated at nearly 45 per cent at the end of 1998, matched this run on the local currency.

Data from the household surveys illustrate how households have responded to increasing economic hardship. Throughout the survey period, wages remained the most important source of income for two-thirds of Chitungwiza households. Trading and vending once mainly informal activities, declined from 17 per cent in 1994 to 11 per cent in 1998. The increasing contribution from 'other sources' in Chitungwiza (nine per cent in 1994 to 25 per cent in 1998) reflects growing diversification of informal activities as a living. In Murehwa, wages are a less important source of income, and reliance on remittances has increased steadily from 27 per cent in 1994 to 48 per cent in 1998, exceeding agricultural production as an income source since 1995. About half of remittances are derived from informal-sector activities and it appears rural men now migrate to seek both formal and informal employment.

The different regular sources of income for households were investigated. The findings suggest a rise in multiple sources of income. In Chitungwiza in 1991, two-thirds of households (67 per cent) had one income source; by 1998 this was down to 32 per cent. Similarly, two per cent had three or more sources of income in 1991, rising to 21 per cent by 1998. Rural Murehwa reported similar levels of income diversification. In 1991, four per cent reported one source of income compared to 28 per cent in 1998. By 1998 less than one-third (31 per cent) of households relied on one income source, compared to 50 per cent in 1991.

Despite increased sources of income, many households failed to earn more money. The real value of the Zimbabwean dollar has declined more than seven-fold against hard currencies since 1988. Despite this decline, about nine per cent of Chitungwiza households earned under ZWD400 per month in 1998 compared to ten per cent in 1995. Likewise in Murehwa, the proportion of families earning under ZWD400 did not change much (46 per cent in 1995 and 36 per cent in 1998). For the poorest families in the lowest income quartile, reported incomes have scarcely changed. At the other extreme, in Chitungwiza, on average, the richest quartile nearly doubled its income in nominal terms since 1995.

Between 1995 and 1998 most households in Murehwa and Chitungwiza reported their income either staying the same or declining. Between one-quarter and one-third of households, usually from higher-income groups, reported an increase in income. Respondents were then asked to compare present income to a year ago. Nearly half (44 per cent) of households in Chitungwiza earning under ZWD400 a month reported income decreasing, whereas the same percentage of households earning ZWD2000 monthly said income was rising. The contrast is more evident in rural Murehwa. Over half of the poorest households said they were earning less, while nearly three-quarters of the wealthiest respondents said they were earning more than the previous year (Table 7.1). In other words, the poorest households most commonly report a decrease in income, whereas the richest households most commonly report an increase. This finding is consistent with growing income inequality in both rural and urban areas.

Finally, economic hardship has had an effect on food security in urban and rural areas. Periodic food shortages affect many households. In Chitungwiza, food availability in 1995 was particularly difficult when 41 per cent of households reported that they did not have enough food to feed their family. In subsequent years, urban food security appears to have improved (12 per cent of households in 1998), although households lost to follow-up study (40 per cent over the six years) may bias these results. In Murehwa district, the situation has worsened each year. In 1994, ten per cent reported periodic or sustained food shortages, rising to 21 per cent and 29 per cent respectively in 1998. Especially worrying is that around 20 per cent of households in Murehwa district in 1996 and 1998 claimed insufficient maize meal. Self-sufficiency in maize fell from 76 per cent to 49 per cent.

Another effect of ESAP was access to health care. Government commitment to maintaining mass access to health services was beyond question in the 1980s, supported by sustained increases in public spending in real terms. This was not continued under structural adjustment. By the mid 1990s, a rising share of public expenditure was needed to repay foreign debts, reaching 25 per cent of government spending in 1996–97. During the 1980s and early 1990s around six per cent of public expenditure was

Table 7.1 Reported change in household income by income category in Chitungwiza and Murehwa district

	Decrease (%)			No change (%)			Increase (%)		
	1995	1996	1998	1995	1996	1998	1995	1996	1998
Chitungwiza									
less than ZWD 400	48	58	44	32	29	28	20	13	28
ZWD 400 to 999	45	32	56	42	46	22	13	22	22
ZWD 1000 to 1999	26	21	54	42	43	18	33	36	28
ZWD 2000 or more	12	12	27	43	37	29	45	51	44
Murehwa district									
less than ZWD 400	54	35	70	39	43	19	8	22	11
ZWD 400 to 999	40	25	60	41	39	21	18	36	19
ZWD 1000 to 1999	31	23	51	34	29	10	34	49	39
ZWD 2000 or more	13	4	24	38	13	4	50	83	72

on health, peaking at nearly seven per cent in 1991. Since 1994–95 this figure has fallen to four per cent, equivalent to a 40 per cent decline per capita and coinciding with a sharp rise in demand for health care, in part due to the growing HIV/AIDS epidemic.

Cost recovery remains one of the most visible components of ESAP. In 1991 the Zimbabwean government began to enforce the collection of user fees for health services, a policy introduced in 1985. Since the policy was adopted, the government had changed its position several times. New regulations were announced in November 1992, raising the income level for eligibility for free treatment from ZWD150 to ZWD400 per month. In January 1993, the government abolished fees at RHCs and most rural hospitals, in order to alleviate the effects of the 1991–92 drought. This relief was not implemented effectively, however, and most council and mission clinics continued to charge fees. In June 1993 the government reintroduced user fees at rural government health facilities and in January 1994 new fee schedules posted increases of, on average, two and a half times. Finally, the government abolished fees at RHCs in March 1995. No further changes have since occurred.

The impact on health-seeking behaviour has been worrying. Since 1994, about one-third of respondents reported an illness 'serious enough to interfere with daily activities' in the previous four weeks. The vast majority reported that they obtained care from the formal health care delivery system, but in both study areas a growing proportion of respondents reported not seeking treatment. In the rural areas in 1998, about one in three cases went untreated, compared to one in five cases in 1993 and 1994. One worrying indicator of reduced utilization of health care is the national immunization coverage rate. This rate peaked for most of the six common childhood vaccines in 1991 but has declined sharply thereafter. For example, measles coverage fell from 90 per cent to 78 per cent between 1991 and 1993 and has continued to remain well below 1991 levels (DHS, 1998).

The effect of economic hardship on access to health care is further illustrated by Figure 7.1, which shows trends in the number of new patients who visited the outpatient departments (OPD) of Murehwa district hospital, the mission hospital and the average rural health centre (RHC) over the study period. It shows that attendance levels at RHCs increased in the first half of 1993 when no fees were charged. There were actually more than twice as many patients seen as in late 1992. The drought and free distribution of food and meals for children under five attracted more patients, but these events covered longer periods than the six months during which there were no fees. The higher level of OPD attendance coincided with this period. Between mid-1993 and early 1995, attendance levels were slightly higher than in 1992. In the second half of 1995 outpatient visits to RHCs rose by more than one-third. Again this is likely to be attributable to the abolition of fees in March 1995.

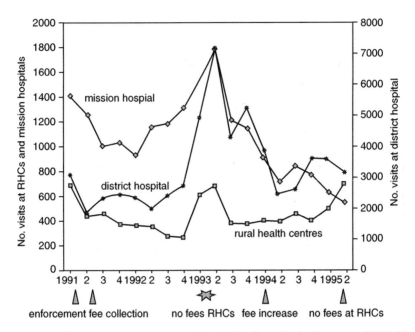

Figure 7.1   First outpatient department visits in Murehwa district by quarter (1991–95)

At the district hospital, attendance levels were also high in early 1993 as a combined result of drought and free health services. During that time, waiting times for patients at the hospital OPD were excessively long. Attendance levels started to fall in late 1993, before the large fee increase of January 1994. The second half of 1994 then showed a slight recovery. At the mission hospital, attendance levels fell during the whole of 1991 and into 1992, when user fees were increased twice. Attendance increased during the 1992 drought year, with a peak in early 1993, similar to what happened at the district hospital. After mid-1993 attendance has been falling almost continuously to reach a level below that of the average RHC. This is remarkable because the mission hospital used to receive two to three times more patients than the average RHC. There is a clear association with changes in user fee policies: the enforcement of fee collection in 1991, the temporary abolition of fees at RHCs in 1993, the January 1994 fee increase and the abolition of fees at RHCs in March 1995.

Similar trends are observable in maternity services. In Chitungwiza the proportion of women paying maternity fees decreased over the years to about two-thirds in 1995. In the rural area, this has not been the case. In the 1995 survey 97 per cent of women who delivered at health facilities reported paying fees. Those who paid in Chitungwiza saw the fees almost double between 1993–94 to ZWD120 and more recently to ZWD200.

Compared to five years previously, this increase was up to eight-fold. The fee increase in Murehwa district was more modest, with most women paying ZWD10–12, although hospital deliveries cost ZWD60. Overall, the majority of women who had a baby between mid-1994 and mid-1995 reported booking for delivery (and paying a fee) when going for antenatal care (93 per cent in Chitungwiza; 90 per cent in Murehwa district). The impact on health care utilization can be seen in the rise of home deliveries in Murehwa district during the same period. The proportion of home deliveries increased steadily from an estimated 18 per cent in 1988–91 to 38 per cent in 1995. After mid-1995, when the government abolished fees at RHCs, the proportion of home deliveries fell back to 17 per cent. In Chitungwiza there is no such trend observed, although a record ten per cent of reported deliveries in 1994 did not take place at a health institution.

7.5   Globalization, structural adjustment and health

The 1990s were a period of economic hardship accompanied by declining health indicators for certain population groups in SSA. It is difficult to attribute the roots of these negative trends – rising infant and child mortality, doubling of maternal mortality and rise in acute malnutrition – wholly to structural adjustment. The devastating drought of 1991–92 plunged the southern African region into recession. Zimbabwe has a diversified economy, but about one-half of GDP is dependent on agriculture. The 1990s also saw the explosion of the HIV/AIDS epidemic, and Zimbabwe now has one of the most severe epidemics in the world (UNAIDS, 2000b).

Yet these difficult events have also taken place amidst macroeconomic policy reforms that have reduced the ability of governments to respond effectively to them. Pressures on public spending and increased servicing of foreign debt impacted adversely on Zimbabwe's capacity to relieve the human effects of drought and disease. By the end of 1998, the government devoted 30 per cent of annual expenditure to servicing the national debt, nearly eight times the entire public budget for health. Despite optimistic predictions of the benefits ESAP would bring, the economy has since been plagued by high inflation, high interest rates, an unstable exchange rate, growing unemployment and contraction of traditionally strong industries such as textiles and steel. There is general agreement that adjustment has failed.

This research provides strong evidence that there has been serious economic degradation of poor households in urban and rural areas of Zimbabwe and there is no sign that this process has ended. The gap between rich and poor has widened and a substantial proportion of the population now experiences severe household food insecurity and dietary deterioration. The government's failure to protect the health sector from budget cutbacks, and to guarantee high quality and affordable services at the primary care level since the introduction of ESAP, has had a negative impact on household welfare. These findings from Zimbabwe are supported by a review of existing studies of SAPs and health undertaken by the WHO Commission on Macroeconomics and Health which states that 'The majority of studies in Africa, whether theoretical or empirical, are negative towards structural adjustment and its effects on health outcomes' (Breman and Shelton, 2001, p. 15). Similarly, there is evidence that SAPs continue to adversely affect the performance of health systems through chronic underfunding of infrastructure, and reduced numbers and quality of health personnel (Simms *et al.*, 2001).

Perhaps most worryingly, SAPs have been joined by other policy instruments of globalization that are likely to further undermine the ability of low-income countries to provide appropriate health care for their populations. For example, multilateral agreements under the World Trade Organization (WTO), notably the Agreement on Trade-Related Intellectual Property Rights (TRIPs) and General Agreement on Trade in Services (GATS), have already threatened to circumscribe the policy options of many countries. Notwithstanding the important change in stance by pharmaceutical companies in relation to HIV/AIDS drugs (see Koivusalo and Thomas chapters in this volume), the provisions of the WTO raise many concerns about how they will impact on the broad determinants of health in low-income countries and particularly the poor (Hong, 2000). As President Museveni of Uganda states, 'It (globalization) is the same old order with new means of control, new means of oppression, new means of marginalisation' (*SA Business Day*, 23 August 2000).

7.6   Conclusions: towards a response to the challenge of Health for All

In recognition of the growing global health divide between rich and poor, compounded by the crises imposed by HIV/AIDS, resurgence of TB and malaria, and other growing challenges, the UN Secretary General recently announced the establishment of a Global Fund to fight AIDS, Tuberculosis and Malaria. Many high-income countries have already contributed to the fund and the initiative is highly welcome by a sector habitually short of resources. However, there is also concern that the lessons of the past two decades described above, and in particular the central importance of PHC to achieving HFA, are not lost in forthcoming discussions on the use of these new and substantial resources. As a recent editorial in the *BMJ* (2001: 1321) states, 'The dominant fear ... was that this new public–private partnership fund would (yet again) be donor led. As a result undue emphasis would be put on supplying drugs rather than building up capacity to implement and sustain effective treatment and preventive programmes.' This view is echoed by a recent review of health systems in Africa that concludes that 'Programmes to tackle these important diseases will not be sustainable in the long-run unless effective health services are in place. International aid should therefore support system development and improve the delivery of health services' (Simms *et al.*, 2001, p. 1).

The above concerns are reminiscent of the familiar, yet relevant, debate between comprehensive and selective PHC. It is clear from the experiences of the past 20 years that a sustainable response to the challenges of HFA in low-income countries must hinge on strategies that address the strengthening of health systems as a whole. Countries that have achieved the greatest, and most durable, improvements in health tend to be those with a commitment to equity and to health systems that are comprehensive and ready to engage related sectors. Strong empirical evidence for this comes from a number of countries, including some of the poorest countries – Sri Lanka, China, Costa Rica and Kerala State in India. These countries demonstrate that investment in the social sectors, and particularly in women's education, health and welfare, has a significant positive impact on the health and social indicators of the whole population (Halstead *et al.*, 1985).

Of particular relevance to the development of comprehensive health systems is the clause in the Declaration of Alma-Ata stating that PHC 'addresses the main health problems in the community, providing promotive, preventive, curative and rehabilitative services accordingly' (WHO and UNICEF, 1978). Comprehensive health systems and programmes include, therefore, curative and rehabilitative components to address the effects of health problems, a preventive component to address the immediate and underlying causative factors operating at the level of the individual, and a promotive component addressing the more basic (intersectoral) causes at the level of society (Sanders, 1999).

The more effective programmes have taken the approach of involving health workers, workers from other sectors and the community in the three phases of programme development, namely, assessment of the nature and extent of the problem, analysis of its multilevel causation and action to address the linked causes. Clearly, the specific combination of actions making up a comprehensive programme will vary from situation to situation. The inclusion of a set of well-tested health service activities should constitute the core of a comprehensive control strategy, for example DOTS for TB, early treatment of STDs and promotion of condom usage for HIV, and effective prophylaxis and treatment and impregnated bednets for malaria. However, these activities, in order to be sustained, need to be embedded within functioning health systems and complemented by relevant promotive activities in health related sectors (such as improved housing and nutrition for TB, lifeskills education for HIV prevention and environmental improvements for malaria).

This shift in focus from selective disease-specific interventions to a more comprehensive health systems approach implies a shift both in policy emphasis, time horizons and scale and duration of investment. Policy needs to recognize the centrality of processes in health systems development and not only the short-term easily measurable outcomes achieved through vertical interventions, and that this requires both time and sustained investment.

To secure sustained investment in the health and social sectors and the equity essential for a healthy society, evidence suggests that a strong, organized demand for government responsiveness and accountability to social needs is crucial (Mosley, 1985). Tacit recognition of this important dynamic informed the Alma Ata call for strong community participation. To achieve and sustain the political will to meet all people's basic needs, and to regulate the activities of the private sector, a process of participatory democracy – or at least a well-informed movement of civil society – is essential. 'Strong' community participation is important not only in securing greater government responsiveness to social needs, but also in providing an active, conscious and organized population so critical to the design, implementation and sustainability of comprehensive health systems.

It is clear that the announcement of the Global Health Fund presents to developing countries an opportunity to mount a response to the health crisis they are facing. However, unless these resources contribute to the development of infrastructure, human capacity and management processes, this response is likely to have only a short-term impact on health problems which are manifestations of economic and social under-development and dysfunctional systems, which, it has been argued, are themselves liable to be further aggravated by globalization.

In conclusion, this chapter reviews the uneven progress towards achieving HFA, particularly in SSA, over recent decades. It argues that the forces of globalization, and particularly SAPs, have had continuing negative impacts on poor families and on their social safety net, including basic health care. Additionally, contemporary instruments of globalization, particularly TRIPS and GATS, threaten to further undermine the capacity of poor governments to adequately serve the social and health needs of the majority of their populations. Using historical and empirical evidence, it is argued that only a broad-based approach to health development, with sustained investment in the social sector and comprehensive health systems, can effectively respond to this crisis. Furthermore, strong community participation, or well-organized and active civil society, seem key to the achievement of equitable social policies and effectively functioning health systems. Thus, while public–private partnerships such as the Global Health Fund are currently being promoted as a response to the crisis in health, this chapter suggests that basic needs, including effective health systems, would be better guaranteed through investment in strong partnerships between governments and civil society.

Notes

1. This section is based on D. Sanders, 'PHC 21 – Everybody's Business', Main background paper for the Meeting: PHC 21 – Everybody's Business, An International Meeting to celebrate 20 years after Alma-Ata, Almaty, Kazakhstan, 27–28 November 1998, WHO Report WHO/EIP/OSD/00.7.
2. This section draws heavily on L. Bijlmakers, M.T. Bassett and D. Sanders, 1999, 'Socio-economic stress, health and child nutritional status at a time of economic structural adjustment – a six year longitudinal study in Zimbabwe', unpublished report, and M. Bassett, L. Bijlmakers and D. Sanders, 'Experiencing Structural Adjustment in Urban and Rural Households of Zimbabwe' in *African Women's Health*, M. Turshen (ed.), Africa World Press, Inc. 2000, pp. 167–91.
3. Gini coefficients (named after the Italian statistician who developed the measure in 1912) are aggregate measures of inequality. The measure can vary from 0 (perfect equality) to 1 (perfect inequality). The Gini coefficient for a country with a highly unequal income distribution typically lies between 0.50 and 0.70. A country with a relatively equal distribution measures about 0.20 to 0.35. For a discussion see Michael P. Todaro, *Economic Development*, 5th edn 1994, (New York: Longman).

Part II

Towards Global Governance for Health

8
Globalization and Cholera: Implications for Global Governance
Kelley Lee and Richard Dodgson

8.1 Introduction

Plague and pestilence have become an increasingly popular theme since the end of the Cold War among policymakers, journalists, fiction writers and film directors searching for new threats to personal and national security. Ill health and, in particular, infectious diseases have generated a spate of popular, and often alarmist, literature (Garrett, 1994; Preston, 1994; Ryan, 1996). This has been accompanied by growing high-level concern within governments and the medical community with global health issues that threaten national interests (Institute of Medicine, 1992, 1997; US Committee on International Science, Engineering and Technology Policy, 1995). The emphasis in many of these discussions has been on emerging health threats that are perceived to pose potentially sudden and serious dangers to public health.

We begin this chapter with the premise that the process of globalization has particular impacts on health and that there is a clear need to better understand and more effectively respond to these impacts. However, without underplaying the dangers posed by health emergencies caused, for example, by the genetic mutation of viral agents or epidemics of emerging infectious diseases, we seek to develop a broader understanding of the historical and structural factors behind the health challenges posed by globalization. As we discuss later, globalization can be defined as a process that is changing the nature of human interaction within a range of social spheres. Globalization's impact on health can be seen as part of a longer historical process firmly located in social change over decades, and perhaps centuries, rather than recent years.

From this perspective, an understanding of global health issues at the turn of the twenty-first century could benefit substantially from the voluminous literature on globalization from international relations, including the subfields of social and political theory and international political economy. This is a rich and highly relevant literature. It documents what structural

changes are occurring toward a global political economy, how power relationships are embedded within this process of change, what varying
impacts this may have on individuals and groups and to what extent global
governance could effectively mediate this process. These issues counterbal-
ance the strong focus in the health literature on biomedical research, information systems and other technical solutions. Although health is a classic
transborder issue, it continues to receive limited attention in international
relations.

We seek to bring together the international relations and health fields for
two purposes. First, knowledge of the globalization process can be used to
better understand the nature of health issues and the development of effective responses to them. To explore this link, we analyse cholera from the
nineteenth century to the present, with particular attention to the seventh
pandemic (1961–present). We argue that the particular form that globalization takes has created social conditions that have influenced the transmission, incidence and vulnerability of different individuals and groups to the
disease. Thus, we compare epidemiological patterns of the disease alongside changing patterns of human migration, transportation and trade.

Second, knowledge of health and disease can be used to better understand
the nature of globalization, because globalization is a highly contested concept, infused with embedded interests and having both positive and negative
consequences (Amoore *et al.*, 1997). Cholera has mirrored this process, high-
lighting the contradictions of globalization in its present form. A significant
and often overlooked threat to human health, therefore, is the particular
form globalization took in the late twentieth century. Thriving in the midst
of increased poverty, widening inequalities within and across countries, pressures to shrink the public sector and global environmental change, cholera
can be seen as a reflection of the ills of globalization itself.

From analysis of this dual relationship between globalization and cholera,
we conclude by considering the implications for existing mechanisms of
international health co-operation. Following a brief review of measures for
the transborder prevention, control and treatment of infectious diseases, we
explore the need for a system of global governance. We propose a definition
of global governance for health and discuss key functions and characteristics that may be needed to protect human health on a global scale.

8.2   Globalization and health: a conceptual framework

An understanding of the linkages between globalization and health depends
foremost on one's definition of globalization and precise dating of the
process. In this study, we define globalization as a process that is changing
the nature of human interaction across a range of social spheres, including
the economic, political, social, technological and environmental. This
process is globalizing in the sense that many boundaries hitherto separating

human interaction are being increasingly eroded. These boundaries – spatial, temporal and cognitive – can be described as the dimensions of globalization (Lee, 2000).

The *spatial dimension* concerns change to how we experience and perceive physical space. Roland Robertson (1992) writes of 'a sense of the world as a single place' because of increased travel, communication and other shared experiences. Conversely, this 'death of distance' (Cairncross, 1997) has also led to more localized, nationalized or regionalized feelings of spatial identity. As such, globalization can be seen as a reterritorializing rather than deterritorializing process. Second, the *temporal dimension* concerns change to the actual and perceived time in which human activity occurs, generally toward accelerated time frames. A good example is currency trading of US$1.7 trillion worldwide each day, two-thirds of this amount retraded after less than seven days. The speed of communication (facsimile, e-mail) and transportation (high-speed train, Concorde jet) have also accelerated social interaction. Third, globalization has a *cognitive dimension* that affects the creation and exchange of knowledge, ideas, beliefs, values, cultural identities and other thought processes. Change has been facilitated by communication and transportation technologies that have enabled people to interact more intensely with others around the world. The production of knowledge has also become more globalized through research and development, mass media, education and management practices. Contrasting forces are at play that, on the one hand, homogenize cognitive processes for better or worse (for example, global teenager) and, on the other hand, encourage greater heterogeneity (for example, religious fundamentalism).

In addition to the precise nature of globalization, the timing of the process has been subject to dispute. Some believe that it is a relatively recent phenomenon defined foremost by the activities of multinational corporations and their striving for global economies of scale (Rugman and Gestrin, 1997). While others agree that social relations have become more intense during the past ten to 20 years, they argue that such relations are not fundamentally new. Anthony Giddens and Roland Robertson, for example, argue that globalization has historical roots from the fifteenth century (Giddens, 1990). Giddens asserts that, since the fifteenth century, globalization has developed with modernity. Robertson, however, disputes the view that globalization has followed a single *telos* or that its emergence can be linked to a single force such as modernity. Instead, he argues that there are a number of historical stages to globalization (Table 8.1), each characterized by what are currently regarded as features of a modern society and global system. These include, for instance, human migration, a system of global trade, urbanization and the growth of international governance.

It is this conceptualization of globalization occurring across multiple spheres and dimensions, and a time frame of centuries, that we adopt in this essay as a useful framework for understanding its complex impacts on

Table 8.1   Robertson's historical stages of globalization

Stage	Time scale	Spatial centre	Characteristics
Germinal	Circa 1500–1850s	Europe	Growth of national community, accentuation of the concept of the individual, spread of Gregorian calendar
Incipient	Circa 1850–1870s	Mainly Europe	Shift toward homogeneous unity of the state, formalized international relations
Takeoff	Circa 1870–1920s	Increasingly global	Inclusion of non-European states into international society, World War I, League of Nations
Struggle for hegemony	Circa 1920–1965	Global	Wars or disputes about the shape of the globalization process, atomic bomb, UN
Uncertainty	Circa 1965–present	Global	Inclusion of Third World, moon landing, end of Cold War, environment, HIV/AIDS

Source: R. Robertson, *Globalization: Social theory and global culture* (London: Sage, 1992).

health. Briefly, the geographical spread of disease can be closely correlated with the migration of the human species across the globe. Robert Clark (1997) writes, 'In becoming global, humans, plants, animals, and diseases have coevolved; i.e., evolved together as a package of interdependent life systems.' With the spread of human populations came changes to social organization into larger communities with different lifestyles, notably from hunting and gathering to agriculture and animal husbandry, to sustain such communities. Changes to disease patterns followed – increased zoonosis (for example, tuberculosis, rabies, salmonella, helminths), nutritional ills and dental decay. The establishment of permanent communities, accompanied by systems of irrigation and often ineffective sanitation and water supplies, led to increased diseases such as malaria and schistosomiasis (Cohen, 1989; Karlen, 1995).

As J.N. Hays points out, human settlements on different continents remained relatively isolated until the late fifteenth century when the age of exploration brought Europe into contact with the Americas. Coinciding with a greater concentration of human populations into larger communities, the impact of the spatial dimension of globalization on health became increasingly evident. Periodically, epidemics could now become pandemics through intercontinental trade, migration and imperialism. During this period, diseases such as plague and influenza travelled the silk route from Asia into Europe, or by ship to ports throughout the old and new worlds.

Similarly, typhus from Asia and syphilis from the Americas were brought to Europe, while Europeans introduced measles, smallpox, typhus, plague and other diseases to the new world, thus precipitating 'the greatest demographic disaster in history' (Hays, 1998).

The greater frequency and intensity of human interaction across continents eventually made the temporal dimension of globalization more prominent. From the seventeenth century, industrialization, rapid urbanization, military conflict and imperialism brought increased vulnerability to many populations. Coupled with inequalities in living standards and a lack of basic medical knowledge, many communicable diseases (such as syphilis, typhus, tuberculosis and influenza) spread more rapidly than ever before. Another disease, the infamous 'potato blight', was also able to travel quickly from the northeast of North America in 1840 to Europe by ship in 1845, eventually causing widespread starvation, notably in Ireland (Cohen, 1995).

By the late eighteenth century, these changes to the spatial and temporal dimensions of health risks and determinants led to greater efforts to develop public health knowledge and practice to respond to them. The nineteenth century brought the initiation of bilateral and regional health agreements, giving way from 1851 to periodic international sanitary conferences to promote intergovernmental co-operation on infectious disease control. This was accompanied by major advances in medical knowledge, including vaccination and microbiology, which were increasingly shared at international scientific meetings. This process of knowledge creation and application across countries signalled the cognitive dimension of globalization in health, which gained further momentum with the professionalization of the health field, creation of research and training institutions worldwide, and growth of scientific publications. Yet, as we show later, how health and ill health within and across societies were understood cannot be separated from prevailing beliefs and values characterizing these later stages of globalization. The creation of the World Health Organization (WHO) in 1948 was fuelled by the postwar faith in scientific and technical solutions to defeat ill health, a belief that would be seriously shaken by the uncertainty of this phase of globalization into the twenty-first century.

8.3 Cholera in the time of globalization: the first six pandemics

We argue that an analysis of cholera from the nineteenth century can offer important insights into the nature of globalization and the specific challenges it poses for human health. Cholera is caused by the ingestion of an infectious dose of a particular serogroup of the *Vibrio cholerae* bacterium.[1] The bacterium is usually taken into the body through contaminated water or food, then attaches itself to the lining of the human bowel and produces a poison (enterotoxin). The infection is often mild or without symptoms, but approximately one in 20 cases is severe. In severe cases, cholera is an acute illness characterized by repeated vomiting and profuse watery diarrhoea, resulting in rapid loss of body fluids and salts. During this stage, the disease is highly infectious through further contamination of local water, soil and food. Without treatment, this can lead to severe dehydration, circulatory collapse and death within hours (WHO, 1993).

Medical historians have long recognized that the epidemiology of cholera has been intimately linked to social, economic and political change. Cholera had been confined for centuries to the riverine areas of the Indian subcontinent, with occasional appearances along China's coast introduced by trading ships and in the Middle East transported by pilgrims travelling to Mecca. The pattern of the disease changed dramatically, however, in the early nineteenth century with the first of six pandemics over the next hundred years (Table 8.2). The intensification of human interaction during this period through imperialism, trade, military conflict and migration (for example slave trade) was a significant factor. The first pandemic occurred between 1817 and 1823, with cholera suddenly moving far beyond its historical boundaries. A number of changes contributed to its spread in that period: the movement of British troops and camp followers throughout the region; construction of irrigation canals with insufficient drainage ditches to raise cash crops; building of a national railway system; impoverishment of rural people by land reforms and taxation; and mass migration as a result of economic hardship. For the first time, cholera became endemic throughout South Asia, from which it was then transported to the Far and Middle East via burgeoning trade links (such as tea, opium), religious pilgrimage and notably military expansionism.

The geographical pattern of cholera during the next five pandemics continued to mirror human activity, spreading from country to neighbouring country. Corresponding with the intensification of links between Asia, Europe and the Americas, cholera became a worldwide disease in 1826. From India, the disease moved beyond Asia to Europe, the Americas and to a lesser extent Africa; it travelled via immigration, troop movements during times of war and peace (for example the Crimean War) and the slave trade. A particularly important factor was immigration from Europe to North America. Cholera first arrived in New York in June 1832 via immigrants from Dublin who, in turn, travelled inland via wagon train. This extended the second pandemic across North America and then south to the Caribbean and South America. Immigrants then became a repeated source of reinfection, facilitated by the building of the intercontinental railway (Speck, 1993).

This close link between epidemiology and mode of transport can be observed throughout the history of the first six cholera pandemics. As well as extending the geographical incidence of the disease, the speed at which it spread also corresponded with prevailing technology. Until the twentieth century, cholera was limited to travel by land and sea. Compared to the

Table 8.2 First six cholera pandemics, 1817–1923

Date	Geographical pattern of spread
1817–23 (6 years)	India (1817) Ceylon, Burma, Siam, Malacca, Singapore (1818–20) Java, Batavia, China, Persia (1821) Egypt, Astrakhan, Caspian Sea, Syria (1823)
1826–38 (12 years)	India (1826) Persia, Southern Russia (1829) Northern Russia, Bulgaria (1830) Poland, Germany, Austria, England, Mecca, Turkey, Egypt (1831) Sweden, France, Scotland, Ireland, Canada, United States (1832) Spain, Portugal, Mexico, Cuba, Caribbean, Latin America (1833) Italy (1835)
1839–55 (16 years)	India, Afghanistan (1839) China (1840) Persia, Central Asia (1844–45) Arabian Coast, Caspian and Black Seas, Turkey, Greece (1846–47) Arabia, Poland, Sweden, Germany, Holland, England, Scotland, United States, Canada, Mexico, Caribbean, Latin America (1848) France, Spain, Portugal, Italy, North Africa (1850)
1863–74 (11 years)	India (1863) Mecca, Turkey, Mediterranean (1865) Northern Europe, North America, South America (1866–67) West Africa (1868) – limited
1881–96 (15 years)	India (1881) Egypt (1883) North Africa, Southern Mediterranean, Russia China, Japan United States, Latin America (1887) – limited Germany (1892)
1899–23 (24 years)	India (1899) Near and Far East Egypt, Russia, Balkan Peninsula Southern Europe, Hungary China, Japan, Korea, Philippines

Note: Accurate historical data on the epidemiology of specific infectious diseases is notoriously incomplete, given the absence of standard reporting systems (for example lists of notifiable diseases), incompleteness of demographic data, and differences in disease nomenclature. This table draws on the limited data available to illustrate the general geographic pattern of early cholera pandemics. The spread of the disease by year and country cannot always be provided comprehensively.

Sources: Compiled from Barua (1972) and Kiple (1993).

seventh pandemic, which we discuss later, the rate of spread was relatively slow. For example, it took the disease three months to cross the sea via trade routes from Hamburg (where it arrived in August 1831) to Sunderland in the northeast of England (where it arrived in October 1831). During later pandemics, the introduction of the steamship led to cholera spreading more quickly across major bodies of water. Hence, in 1848 cholera advanced from Poland to New Orleans in just over seven months. This more rapid move also coincided with the approximate doubling of railway lines and tonnage of steamships in operation (Hobsbawm, 1975), correlating with the rapid spread of the disease throughout the Americas. By the time of the third and fourth pandemics, the spread of cholera into South America was hastened by the opening of the Panama Canal. A similar pattern can be observed in the Middle East with the opening of the Suez Canal.

Another feature during this period was cholera's close association with prevailing social conditions that enabled the disease to spread so widely and repeatedly, in the process becoming endemic in many parts of the world. In India, the cumulative effect of profound changes to local societies and ecology was, as Sheldon Watts writes, the transformation of a merely local disease, endemic in Bengal, into a chronic India-wide problem by 1817 and soon afterward an epidemic disease of worldwide proportions (Watts, 1997). As cholera spread, it found fertile conditions for epidemic transmission in nineteenth-century industrialization. Poor sanitation, poverty, malnutrition, overcrowding, ignorance and a lack of basic health services allowed cholera to flourish in the new urban centres of Europe and North America. Such was the link between cholera and social inequality during this period that the disease added impetus to fermenting unrest. In Russia, an uprising by revolutionaries known as the 'cholerics' in the 1830s was because of the belief that the disease was actually a plot to kill off the poor. Although Tzar Nicholas I quelled the movement, cholera continued to inflame class conflict across Europe.

By the middle of the nineteenth century, health policies to control epidemic diseases began to be adopted. But the link between disease and the squalid conditions of the poor had not yet been made officially. The experiences of Florence Nightingale with cholera during the Crimean War led the British Army to establish sanitary engineering as a new branch (Bray, 1996). The first UK Public Health Act was adopted in 1848, followed by the creation of a General Board of Health, although cholera continued to be blamed on immorality and a lack of 'proper habits' (Watts, 1997, p. 194). Despite John Snow's historic gesture in 1854,[2] it was not until the discovery by Robert Koch in 1883 of the bacillus *Vibrio cholerae* as the causative agent of cholera that the relative inertia of government bodies toward the lack of safe drinking water and sanitation for poor people finally ended. This was gradually followed by the establishment of a public health infrastructure, eventually supported by a system of regulation and social welfare.

Internationally, governments complemented national efforts with meetings to promote co-operation and improve public sanitation. Between 1851 and 1911, there were 12 International Sanitary Conferences, which led to the formation of international public unions, such as the International Association of Public Baths and Cleanliness. In 1907, the Office International d'Hygiène Publique (OIHP) was created to standardize surveillance and reporting of selected communicable diseases, including cholera. Following establishment of the League of Nations, the work of these international public unions continued under the auspices of the League's Health Commission. In particular, the commission 'established new procedures for combating epidemics and initiated studies in child welfare, public health training and many other subjects' (Armstrong, 1982). All of these efforts proved to be the forerunners of increasingly organized international health co-operation, leading to the eventual creation of the World Health Organization in 1948.

In summary, the first six pandemics of cholera can be understood in close relation to the prevailing socioeconomic and political structures of the period. From a local disease, cholera became one of the most widespread and deadly diseases of the nineteenth century, killing an estimated tens of millions of people (Bray, 1996). Not coincidentally, cholera travelled the same routes around the globe as European imperialism. The disease, in this sense, was an integral part of this stage of the globalization process, affected by changing spatial, temporal and cognitive dimensions of human interaction, but, in turn, also influencing the articulation or particular form that globalization has taken. It is within this historical context that the seventh pandemic can be understood.

8.4 Globalization and the seventh pandemic: mirror, mirror on the wall

The seventh cholera pandemic began in 1961 in Sulawesi, Indonesia, after a gap of 38 years. The disease remained endemic in a number of regions, including South and Southeast Asia, and cases were reported regularly until the 1960s. Nonetheless, there was a declining incidence overall, and worldwide transmission did not occur. The primary reasons were improvements in basic sanitation and water supplies in many countries, backed by international health co-operation. Until 1961, it was thought that cholera was disappearing.

But in 1961, cholera presented a new and unexpected challenge (Table 8.3). Unlike previous pandemics, which are believed to have been the result of the classical biotype, the cause of this new pandemic was discovered to be a different biotype of *Vibrio cholerae*, known as El Tor.[3] Although less virulent, this new strain has proven more difficult to eradicate. El Tor cholera causes a higher proportion of asymptomatic infections, allowing carriers to spread the disease through contamination of food or water. It survives

Table 8.3 Seventh and eighth cholera pandemics, 1961–present

Pandemic, date, type[a]	Countries
Seventh pandemic, 1961–present, El Tor biotype	Indonesia (1961)
	Indo-Pakistan Subcontinent (1963–64)
	West Pakistan (Bangladesh), Afghanistan, Iran, Uzbekistan, Thailand (1965)
	Iraq (1966)
	Laos, South Korea, Hong Kong, Macao, Nepal, Malaysia, Burma, East Pakistan (1969)
	Russia (Astrakhan, Odessa/Kersh), Turkey, Czechoslovakia, France, UK,[b] Lebanon, Israel, Syria, Jordan, Libya, Tunisia, Dubai, Kuwait,[b] Saudi Arabia, Somalia, Ethiopia, Guinea, Sierra Leone, Liberia, Ghana, Côte d'Ivoire, Mali, Togo, Dahomey, Upper Volta, Nigeria, Niger, Japan[b] (1970)
	Muscat, Oman, Yemen, Morocco, Algeria, Cameroon, Chad, Mauritania, Senegal, Kenya, Uganda, Madagascar,[b] Spain, Portugal, France,[b] Sweden,[b] West Germany[b] (1971)
	United States (1973)
	Peru, Colombia, Chile, Bolivia, Brazil, Ecuador, Guatemala, Honduras, Mexico, Nicaragua, Panama, Venezuela (1991)
	Argentina, Belize, Costa Rica, El Salvador, French Guyana, Guyana, Surinam (1992)
	Paraguay (1993)
	Zaire, Ukraine (1994)
suspected Eighth pandemic, 1993–present, 0139 Bengal serogroup	India, Bangladesh (1993)
	Pakistan, Thailand (1994)
	10 Southeast Asian countries, United States[b] (1995)

Notes:
[a] Many serogroups of vibrio cholerae have been identified. During the first to seventh pandemics, only the 01 serogroup (of which two biotypes exist) caused cholera. Its classic biotype caused the first six pandemics and its El Tor biotype caused the seventh pandemic. A different serogroup – 0139 Bengal – is possibly causing the eighth pandemic, the first non-01 serogroup to have caused cholera.
[b] Shows imported cases.
Sources: Barua (1972) and Epstein (1995).

longer in the environment and shows greater resistance to antibiotics and chlorine. It can also live in association with certain aquatic plants and animals, making water an important reservoir for infection (WHO, 1993).

From Indonesia, El Tor cholera moved west to reach India and the Middle East by 1966. In 1970, it reached southern Europe (Russia) and north, east and west Africa. By 1971, outbreaks had been reported in 31 countries, one-third of them experiencing the disease for the first time.

In that year, 150,000 cases were reported, including some 50,000 cases in West Bengal refugee camps. In most newly affected countries, the disease caused severe outbreaks with mortality rates of 40 per cent or higher (Bruce-Chwatt, 1973, p. 86). A year later, it reached the southeastern Mediterranean and eastern Europe; M. Narkevich and others describe the pandemic up to the 1990s as falling into three periods: 1961–69, 1970–77 and 1978–89. The peak of the pandemic was during periods 1 and 2 (1967–74), after which morbidity declined until 1985. Then, from 1985 to 1989, the pandemic seemed to accelerate once again, with 52,000 cases reported in 36 countries. In total, between 1961 and 1989, approximately 1.72 million cases of cholera were reported to WHO from 117 countries (see Table 8.4) (Narkevich *et al.*, 1993).

Explanations for the origins of the pandemic were similar to previous ones, namely transmission around the world from an endemic country via travel. As Leonard Bruce-Chwatt observed, the impairment of sanitation on many Indonesian islands because of overpopulation of urban peripheries, military operations and other disturbances, combined with certain cultural habits (for example use of night soil), allowed cholera to reach epidemic proportions (Bruce-Chwatt, 1973). However, the epidemiology of this pandemic has proven different in two fundamental ways: it has been more geographically widespread (spatial dimension) and has lasted longer (temporal dimension). First, the seventh pandemic has encompassed a large number of countries that have either never experienced cholera before, not done so for many decades, or never to such an intensity. In many parts of Africa, in particular, public health officials have experienced difficulty controlling the disease, largely because of a lack of adequate surveillance, treatment or prevention measures. As D. Barua wrote, 'All the factors favouring endemicity of cholera exist in present-day Africa, particularly in the populous coastal and riverine areas, where there is little possibility of improving water supply, waste disposal and personal hygiene in the near future. In all probability, cholera is going to become, if it is not already, entrenched in this continent at least temporarily' (Barua, 1972). Barua's predictions have proven correct, and cholera has now become endemic throughout west, east and southern Africa for the first time (van Bergen, 1996).

Table 8.4   First three periods of the seventh cholera pandemic, 1961–89

Date	Reported cases	Countries reporting
1961–1969	419,968	24, mainly in Asia
1970–1977	706,261	73 (27 Asia, 32 Africa, 12 Europe, 2 Americas)
1978–1989	586,828	83

Source: Adapted from Narkevich *et al.* (1993).

This unprecedented geographical spread is also observable in the Americas, where the majority of cases have occurred in the 1990s. After disappearing from the Western Hemisphere for almost a century, El Tor cholera was simultaneously reported in two cities in Peru in January 1991. By mid-February, there were 12,000 confirmed cases. WHO described how the epidemic moved with 'unexpected speed and intensity', travelling quickly 2000 kms along the coast to Ecuador. By March–April, it had reached Colombia and Chile. By the end of the year, the epidemic had reached a new country every month, resulting in nearly 400,000 cases and more than 4000 deaths. This was more than the total number of reported cases worldwide for the previous five years. The epidemic continued into 1992, with more than 300,000 cases and 2000 deaths in 20 countries (Guerrant, 1994).

The second difference of the seventh pandemic has been its duration so far – 39 years. This has been by far the longest pandemic and it shows few signs of having run its course. As we discussed earlier, cholera took several months to spread from country to neighbouring country during the nineteenth century, transported by land or sea. El Tor cholera has also travelled readily by land and sea. It is believed, for example, that the disease was imported into Turkey and Lebanon in the early 1970s by workers coming from neighbouring affected countries. Similarly, severe outbreaks in Mali, Ghana, Niger, Nigeria and Chad were traced to the arrival of individuals from infected areas. In India, an outbreak in 1971 was primarily among refugees from the former East Pakistan (Barua, 1972, pp. 426–48). Perhaps most dramatically, it is thought that the epidemic in Latin America began after a ship from China emptied its ballast tanks in Peruvian waters. The *Vibrio* then infected local seafood eaten by local people. El Tor is also thought to have arrived on the US Gulf Coast in the hulls of ships from Latin America (McCarthy *et al.*, 1992).

Since the 1960s, however, the reduced cost of transportation and the addition of faster technologies (for example high-speed rail, ocean liners and air travel) have brought unprecedented movement of human populations to and from endemic areas. Since the 1960s, mass travel (growth of 7.5–10 per cent per annum) has been growing at a rate faster than global population growth (growth of 1.5–2.5 per cent per annum) (Cliff and Haggett, 1995). This intensification of human mobility has posed new problems for public health officials, including their capacity to contain serious outbreaks of infectious disease. Kaferstein *et al.* (1997) note that 'over the last two hundred years, the average distance travelled and speed of travel have increased one thousand times, while incubation periods of disease have not.' As the average journey time of an airliner or a bulk carrier is much shorter than the incubation period of a disease, these different forms of transportation are believed responsible for the increased spread of disease from one location to another (Plotkin and Kimball, 1997). Air travel, for instance, brings large numbers of people into close contact with each other within an enclosed space. In relation to cholera, the disease has been spread via contaminated food served on aeroplanes (Heymann and Rodier, 1998) and through global trade in food products (for example shellfish, frozen coconut milk).

Furthermore, the epidemiology of the disease has been geographically unpredictable. Whereas in the past cholera could be monitored from country to neighbouring country, more recently it has 'jumped' continents. In 1973, for instance, 40 passengers travelling from London to Australia were infected with cholera from contaminated food taken on board in Bahrain (Bruce-Chwatt, 1973: 87). Similarly, 31 passengers were found to have cholera on a flight bound for Los Angeles from Buenos Aires, Argentina, in 1992 (Booth, 1991). At least 75 people contracted cholera by eating cold seafood salad loaded on a flight at Lima, Peru, bound for California (Tauxe *et al.*, 1995). It is estimated that, for air travellers from the United States to India, the reported rate of cholera cases is 3.7 cases per 100,000 travellers (Weber *et al.*, 1994). It is this 'hypermobility' of the causative agent via worldwide transportation networks that has been a key factor in the continuation of the pandemic.

Technological change, however, does not tell the entire story of cholera in the late twentieth century. Indeed, like previous pandemics, socioeconomic and political structures have also been central. The 1920s to 1960s were perhaps a 'boom' period for public health systems. Many countries created national health systems. And the aid to the health sector grew through bilateral aid agencies, multilateral organizations (such as WHO, UNICEF), charitable foundations (such as the Rockefeller Foundation) and other nongovernmental organizations (such as Save the Children Fund, Oxfam). Although it was clear that not all benefited equally from this intense period of health development, facilitated by advances in medical knowledge, there was a feeling that progress against many traditional scourges of humankind was being achieved. Eradication campaigns against malaria and (more successfully) smallpox were reflections of this belief in the triumph of science over disease.

The seventh pandemic has been a reminder, however, of the persistent and, in many cases, growing inequalities that remain within and across countries. Life expectancy ranges from 43 years in the poorest countries to an average of 78 years in higher-income countries. Three-fifths of people in the developing world lack access to safe sanitation, one-third to clean water and one-fifth to modern health services of any kind (Crossette, 1998). The initial outbreak of cholera in the early 1960s took advantage of these conditions to spread among the have-nots in many different societies. By the early 1970s, the pandemic seemed to peak and cases gradually decreased over the next decade.

By the 1980s, however, cholera began to benefit from fundamental changes in many public health systems around the world. Despite the launch of the primary health care movement in 1978 by WHO and UNICEF, the 1980s saw a pulling back of the state from the financing and provision of health care. Detailed analysis of this period of health sector reform can be found elsewhere (Ugalde and Jackson, 1995). Briefly, as part of the World Bank's structural adjustment programme, many lower-income countries were encouraged to reduce public expenditure on health throughout the 1980s and 1990s (World Bank, 1987, 1993). Globally, health sector reform in higher- and lower-income countries included policies to encourage market forces to play a greater role in health systems (Ruggie, 1996). There was growing evidence in the late 1990s that these policies have impacted adversely on public health capacity in many countries (Beaglehole and Bonita, 1997; Hoogvelt, 1997), and the growing awareness of globalizing forces has led to efforts to redefine public health functions in countries around the world (Bettcher *et al.*, 1998).

It was in this context that cholera opportunistically spread in the early 1990s. In Latin America, the adverse impacts of globalization on health systems have included increased national debt, rapid urbanization, environmental degradation, inequitable access to health services, and reduced public expenditure on public health infrastructure (Cardelle, 1997). Cholera then arrived in 1991, spreading rapidly across the continent in an epidemic of 1.4 million cases and more than 10,000 deaths in 19 countries (Sanchez and Taylor, 1997). This scenario has not, however, been confined to Latin America. In October 1994, El Tor cholera was reported in the former Soviet Union amid economic instability, deteriorating health services, drought and poor hygiene. The most serious outbreak occurred in ten cities in the Ukraine and threatened a population of 50 million people (WHO, 1996e). As the head of the Ukrainian parliament stated at the time, 'The spread of cholera and other infectious diseases is the calling card of an economy in trouble' (Alexander Moroz, quoted in Ryan, 1996: 108).

Another contributory feature of the global political economy to the seventh pandemic has been the mass migration of people. In the early 1990s, it was estimated that 500 million people crossed international borders on commercial airlines annually. There are 100 million migrants in the world today, with an estimated 70 million people, mostly from low-income countries, working legally or illegally in other countries. In addition, a large proportion (30 million) of total migrants do so involuntarily, including 20 million refugees (Wilson, 1995). Enforced migration has become a particular feature of Africa, where 16 million people have been internally displaced (*New Internationalist*, 1998). In many cases, such migration has been a cause and effect of 'complex emergencies', which have 'a singular ability to erode or destroy the cultural, political, and economic integrity of established societies' (Duffield, 1994).[4] Given large numbers of displaced people, resulting in overcrowded living conditions, poor sanitation, unclean water supplies and malnutrition, cholera has become a familiar feature. For example, the mass population movements as a result of the Pakistani–Indian war in 1971 led to thousands of deaths from cholera and further spread of the disease (Zwi, 1981). Between 1987 and 1991, cholera was diagnosed in Mozambican refugees as they migrated from their home villages to camps in Malawi (Cookson *et al.*, 1998). In August 1994, El Tor cholera broke out in relief camps in Goma, Zaire among Rwandan refugees. Within 24 hours, 800 people had died; the epidemic eventually caused 70,000 cases and 12,000 deaths (Heymann and Rodier, 1998; Sanchez and Taylor, 1997).

Finally, globalization in its present form is believed to be contributing to changes in the natural environment that enable infectious diseases such as cholera to thrive (McMichael *et al.*, 1999). Like the British-built irrigation canals in India, which increased the incidence of malaria and cholera, rapid urbanization and industrialization without sufficient attention to sustainable economic development have led to contamination of drinking water. Historically, the Ganges river has symbolized purification for Hindus, who believe drinking and bathing in its waters will lead to salvation. Today 29 cities, 70 towns and countless villages deposit about 345 million gallons of raw sewage a day directly into the river. Factories add another 70 million gallons of industrial waste and farmers another six million tons of chemical fertilizers and 9000 tons of pesticides (Crossette, 1998). Perhaps more worrisome still, it is believed that widespread changes in coastal ecology are generating 'hot systems' in which mutations of the cholera organism are being selected and amplified under new environmental pressures and then transferred to human populations through the food chain. This is the explanation for the appearance of *Vibrio cholerae* 0139 Bengal in 1992, a new strain of cholera and the first non-01 strain capable of causing epidemics, signalling the beginning of the eighth pandemic (Epstein, 1995).

In summary, the seventh pandemic is a reflection of the contradictions of globalization in the late twentieth century. As in previous pandemics, cholera and deprivation remain closely linked. Development of national health systems worldwide led to the reduction, and to what was hitherto believed to be the eradication, of cholera in many parts of the world. The global spread of cholera since the early 1960s, however, has revealed persistence and growing inequalities within and across countries. Other features of globalization, notably mass migration, social instability and environmental degradation, present the disease with the opportunity to establish itself in new areas of the world. Cholera, in short, is a mirror for understanding the nature of globalization.

At the same time, an understanding of globalization is needed to explain the distinct epidemiological profile of El Tor cholera. Through changes in human interaction spatially, temporally and cognitively, cholera has become more geographically widespread and persistent over time. That the disease can 'jump' across continents within hours, for example, poses new challenges for national health systems. Indeed, the globalization of cholera may require new responses that more closely integrate different levels and types of governance.

8.5   Global governance for health: learning lessons once again from cholera

The emergence of global governance as a central concept in international relations responds to a perceived change in the nature of world politics. In contrast to international governance, the defining feature of global governance is its comprehensiveness. Global governance views the globe as a single place within which the boundaries of the interstate system and nation-state have been eroded. Although the nation-state remains an important actor, processes and mechanisms of global governance are growing to encompass the structures of international governance that manage the system of nation-states. The emerging processes and mechanisms of global governance can be seen as forms of supraterritorial authority (Scholte, 1997).

The processes and mechanisms of global governance are diverse, as are the actors and structures that participate within them. In addition to nation-states, these actors include international institutions, governmental organizations, various nonstate actors, regimes, values and rules (Rosenau, 1995). These different actors compete with each other to shape the nature of global order, the establishment of which is the main purpose of global governance. Global governance may be used to stabilize and expand market capitalism on a global scale, as sought by the International Monetary Fund (IMF) and World Bank, or to establish an order based on greater social justice and redistribution of global resources (Drainville, 1998; Falk, 1995).

Although scholars, practitioners and policymakers in the health field may not explicitly recognize or widely use the term *global governance*, there is growing recognition of the need to establish more effective mechanisms for addressing a range of global health issues. Such issues are wide-ranging and are reviewed elsewhere (Lee, 2000). Developing responses to them leads to questions concerning scope of activity, distribution of authority, decision-making process, institutional structure, and resource mobilization and allocation.

The development of an effective system for the prevention, control and treatment of infectious diseases is perhaps the classic transborder health issue. It has long been recognized that diseases do not respect national borders and that states acting alone are unable to prevent their spread. In recent years, this message has been stated with renewed vigour in relation to emerging and re-emerging diseases and globalization. As a US report warns, 'The modern world is a very small place, where any city in the world is only a plane ride away from any other. Infectious microbes can easily travel across borders with their human and animal hosts ... [And] diseases that arise in other parts of the world are repeatedly introduced into the United States, where they may threaten our national health and security' (US Committee on International Science, Engineering and Technology Policy, 1995). Similarly, Paul Farmer writes that 'EIDs [emerging infectious diseases] have often ignored political boundaries, even though their presence may cause a certain degree of turbulence at national borders. The dynamics of emerging infections will not be captured in national analyses, any more than the diseases are contained by national boundaries' (Farmer, 1996).

A brief review of the existing institutional framework for international co-operation on infectious diseases shows a strong emphasis on biomedical understanding of human disease and state-based health systems. Since its founding in 1948, WHO has carried out surveillance and monitoring of various infectious diseases. Diseases deemed to pose a particular international threat, however, have been governed since 1951 under the International Health Regulations (IHR), a consolidation of various International Sanitary Conventions adopted from the nineteenth century onward. Over the years, the IHR have been periodically updated to take account of changing health needs (for example removal of smallpox from the list of notifiable diseases after eradication). In their present form, the IHR set out procedures for limiting the transmission of infectious disease via shipping, aircraft and other modes of transport. The regulations do not control the movement of international traffic directly but concentrate on controlling the spread of disease where transborder conveyance may occur. The IHR also call on states to 'report to WHO, within specific periods, cases of these three diseases [that is, cholera, plague and yellow fever] within their territories. Second, to facilitate reporting and deter unnecessary interference with international travel and trade, members must limit their responsive health measures (applied to international traffic for the protection of their territories against these diseases) to maximum measures permitted by these regulations' (quoted in Plotkin and Kimball, 1997). Responsibility for the IHR and other infectious disease-related activities falls on WHO's Cluster on Communicable Diseases (CDS), formerly the Division of Emerging and Other Communicable Disease Surveillance and Control (EMC) (now referred to as Communicable Disease Surveillance and Response).

Other international agreements concerned with infectious diseases are the WHO and Food and Agriculture Organization's Codex Alimentarius, the World Trade Organization's Agreement on the Application of Sanitary and Phytosanitary Measures, and the International Civil Aviation Organization's Facilitation to the Convention on International Civil Aviation. Supporting international co-operation, in principle, is a network of national health systems led by ministries of health in each member state and extending to national health services, research institutions, public health laboratories, and monitoring and surveillance systems. Together, it is assumed that WHO coordinates top-down guidance and information, provided from the bottom up by national health systems.

In practice, there are gaps at both the international and national levels. Internationally, the IHR cover only three diseases, and WHO does not have the means to enforce compliance with even these limited stipulations. International surveillance and monitoring relies on the goodwill of governments, but fears of adverse effects on trade or tourism can lead to underreporting. India lost an estimated US$1700 million in exports, tourism and transportation services because of the outbreak of plague in 1996 (Kinnon, 1998). Exports of shellfish by Latin American countries were similarly affected by reports of cholera in the region. Although efforts are in play to revise the IHR to stipulate reporting of 'syndromes' rather than diseases, the question of authority remains.

Another difficulty lies in the limited resources available to mount rapid responses to major transborder outbreaks. Given the limits of WHO resources, the US Centers for Disease Control and Prevention (CDC) at times has stepped in more quickly in health emergencies (such as the outbreak of Ebola in northern Zaire in 1976 and the refugee camps in Goma, Zaire). At the national level, variation in health capacity is even more acute. Many lower-income countries have less than US$4 per capita to spend on health care annually (WHO, 1995). Structural adjustment programmes place further pressure on public health expenditure, and the World Bank continues to attribute problems, such as the re-emergence of cholera, on the failure to privatize rather than on a crisis in resources. As one World Bank study concludes,

> The return of cholera in 1991 to Latin America and [the] Caribbean region was only a symptom of the deep-seated problems and the fragility and inadequacy of publicly operated water supply and sanitation systems. Consequently, the agencies that operate these systems are entering a crucial phase of deciding whether they can greatly improve their operations while remaining in the public sector or whether they should seek increasing private sector financing and participation in both operations.
>
> (Idelovitch and Ringskog, 1995)

To address these deficiencies, international efforts have focused on improving surveillance, monitoring and reporting systems. WHO is now working to create a Global Surveillance Network using electronic links for rapid exchange of information (WHO, 1998d). In 1994, the Program for Monitoring Emerging Diseases (ProMED) was created with the impetus of 60 prominent experts in human, animal and plant health. Accessed via the Internet, ProMED is intended to be a global system of early detection and timely response to disease outbreaks. Similarly, the Global Health Network was established to monitor the spread of emerging infectious diseases through links with public health organizations, multilateral organizations, NGOs and independent research centres. Although such technology is clearly a vital feature of a global system of disease control, this emphasis on technical issues highlights two further flaws in the present system of international health governance.

First, underlying these initiatives is a rational model of policymaking, which assumes that lack of information is the key factor in the globalization of infectious disease. However, as we argue in this chapter, the seventh cholera pandemic has been shaped by the structural features of the global political economy, which have contributed to the vulnerability of certain populations and environments within and across countries. Second, globalization in its present course has created transborder externalities in the form of health risks that increasingly defy state-centric approaches of the past. Although national epidemiological data remain significant, there is also a need for disaggregated data that allow comparisons and analysis within and across countries and regions. Furthermore, focusing on control measures at the national level amid intensified human interaction has so far reinforced a fortress mentality among many governments (Institute of Medicine, 1997). Yet effective control of all infectious diseases at national borders is neither practicable nor ethical. Many diseases remain asymptomatic or have long periods of incubation.

It is here that the existing literature on global governance may offer ideas for creating new forms of authority and institutional linkages to address the challenges of global health. To begin with, the global governance literature highlights the need to be more comprehensive in our approach to global health. Needed, for instance, is a greater appreciation of the link between health and the environment. Douglas Bettcher and Derek Yach point out that public health issues are central components of sustainable development programmes and should be incorporated into the system of global governance that has emerged from the United Nations Conference on the Environment and Development (1992) around sustainable development (Bettcher and Yach, 1998). Piggybacking on strategies for global sustainable development may boost the chances of a system of global health governance being established. Similar linkages between health and changing governance of other sectors, such as agriculture, transportation, communications and trade and finance, need far greater exploration.

Another central theme of the global governance literature is the need to ensure that the processes and mechanisms of global governance have the support of those governed. Various proponents of global governance agree that there is a need to reform the UN's core institutions so that they are more democratic and representative of the global population. WHO, for example, is an organization often described as governed by a 'medical mafia' and influenced by the extrabudgetary funding of a small number of donor governments (Pitt, 1992; Vaughan *et al.*, 1995). At the national and subnational levels, as well, there have long been calls for going beyond the traditional focus on ministries of health and government institutions. Civil society, in particular, is identified as requiring a greater role in policymaking at many levels. For example, the influence of certain global social movements, such as the women's health movement, has been recognized as positively contributing to the democratization of certain policy issues (such as population policy).

The nation-state and existing institutions of international health governance (for example the WHO and IHR) will remain central to any future system of global health governance. The challenge for scholars and policymakers is how to construct processes and mechanisms of global health governance that recognize the interests of nation-states and civil society. This is a massive challenge, yet recent developments in global health governance suggest some optimism about the ability to meet it. For example, plans for a Framework Convention on Tobacco Control bring together shared interests across nation-states, civil society and the business community (for example the pharmaceutical sector) for a more comprehensive effort to control the production and consumption of tobacco. WHO's policy document *Health for All in the 21st Century* adopts a similar approach to deal with persistent inequalities in health within and across countries (WHO, 1998e). These initiatives are directly concerned with strengthening global health governance, proposed by an intergovernmental organization (such as WHO) and supported by an amalgam of member states, civil society groups and individuals.

Finally, there is a need for critical analysis of health determinants that conceptualize social change in the context of long-term and fundamental socioeconomic structures, rather than shorter-term, technically specific change. Such an approach leads to a recognition of the structure–agent links that exist between globalization and health, including the possibility that present forms of globalization are, in fact, incompatible with human health. Importantly, critical theorists see globalization as a historical process constructed not by rationality but by embedded power relations and consequences. Locating cholera and other global health issues within this reflexive starting point therefore seeks to address, and ultimately redress, the underlying causal factors behind ill health.

8.6   Conclusion

In summary, we have analysed cholera from the nineteenth century as a case-study of the links between globalization and health. How human populations have lived – population size and distribution, social structure, cultural practices, distribution of resources – has historically been linked to patterns of health and disease. We have also sought to show how globalization is changing the nature of human societies across the world and, consequently, the health of populations. At the same time, cholera has been reflective of particular features of globalization from its earlier stages to the present.

Globalization has shaped the pattern of the disease, the vulnerability of certain populations and the ability of public health systems to respond effectively. The contributing features of globalization to the epidemiologically distinct seventh pandemic have included socioeconomic instability, intensified human interaction and mobility, environmental degradation and inequalities within and across countries.

As public health systems have struggled to control the seventh pandemic, reports of an eighth pandemic came in 1993. As we described earlier, the possible new pandemic involves a new strain of *Vibrio cholerae*, believed hardier than El Tor cholera in terms of environmental adaptation. Its emergence is thought to have resulted from changes to coastal ecologies in South Asia. By 1995, the pandemic spread to Calcutta, India, with 15,000 cases and 230 deaths, and then moved rapidly to Dhaka, Bangladesh, where 600 cases were reported daily. Severe flooding in Bangladesh in 1998 worsened conditions significantly. Reaching 100,000 cases by 1996, the disease has since spread to Pakistan, Thailand and ten other Southeast Asian countries. Travel-associated cases have been reported in the United States, Europe and Japan (Sanchez and Taylor, 1997). Cholera is poised, it seems, to offer yet another opportunity to learn hard lessons.

Acknowledgement

This chapter was originally published in *Third World Quarterly*, special edition vol. 23(2), April 2002.

Notes

1. There are more than 60 serogroups of *Vibrio cholerae*, but only serogroup 01 causes cholera. Serogroup 01 has two biotypes (classical and El Tor) and each biotype has two serotypes (Ogawa and Inaba).
2. Amid a severe cholera outbreak in London, John Snow found a concentration of cases around a pump in Broad Street. Hoping to prove his theory that cholera is a waterborne disease, he removed the pump handle in front of public officials; and the outbreak was contained.
3. The causative agent is named after the El Tor quarantine camp on the Sinai Peninsula, where it was first isolated in 1905 from the intestines of pilgrims returning from Mecca.
4. Duffield defines complex emergencies as 'protracted political crises resulting from sectarian or predatory indigenous responses to socioeconomic stress and marginalisation'.

9
Antimicrobial Resistance: a Challenge for Global Health Governance

David P. Fidler

9.1 Antimicrobial resistance as a global public health problem

The global infectious disease crisis has many frightening aspects, including the scale of the devastation wrought by the HIV/AIDS pandemic, the emergence of new infections (such as Ebola, Nipah virus, hepatitis C), the bioterrorism threat, and the re-emergence of old killers, such as cholera and tuberculosis. One nightmare cuts across all the grim manifestations of the global infectious disease crisis – antimicrobial resistance (AMR). Antimicrobials are drugs that kill or inhibit the growth of micro organisms (Manivannan and Sawan, 2000). AMR occurs when bacteria, viruses, parasites and fungi develop resistance to antimicrobial drugs, rendering the drugs less effective or useless against the microbes' abilities to cause morbidity and mortality in humans. AMR constitutes one of the most serious global public health problems.

In the twentieth century, antimicrobials became powerful public health weapons in controlling bacterial, viral, parasitic and fungal infections. Today, these weapons are rapidly losing their effectiveness, leading officials at the WHO to warn about the possible return to the pre-antimicrobial age (WHO, 2000b). 'In previous epochs', WHO reminds us, the lack of effective medicines meant that 'a simple bladder infection could lead to death by kidney failure; minor skin conditions such as impetigo could end in scarring and lifelong disfigurement; and killers such as measles, tuberculosis and pneumonia stalked uncontested through the streets, offices and homes of every city and hamlet around the world' (WHO, 2000c).

AMR adversely affects efforts to deal with five of the world's most serious infectious disease killers – HIV/AIDS, tuberculosis, malaria, pneumonia and diarrhoeal diseases (for example cholera, shigella and typhoid) (WHO,

1999b). In high-income countries where antiretroviral drugs have been available, experts are seeing resistance to zidovudine (AZT) and protease inhibitors (Kuritzkes, 2000; WHO, 2000c), which threatens to undermine the treatment options these drugs have presented through expensive research and development. With the global push to make antiretroviral drugs available in low-income countries where HIV/AIDS is most devastating (UNAIDS Expert Group, 2001) antiretroviral resistance may go global. HIV/AIDS also contains an AMR multiplier effect in that it attacks the immune system, making the infected individual vulnerable to secondary infections, such as tuberculosis and pneumonia, which themselves show resistance to traditional antimicrobial treatments.

Pneumonia, as an acute respiratory infection, is one of the world's leading causes of morbidity and mortality (WHO, 1999b). Most of the illness and death that pneumonia causes occur in low-income countries where lack of primary health care and poverty create fertile conditions for infection (WHO, 2000c). WHO reports that '[i]n lab samples as many as 70% of chest infections are resistant to one of the first-line antimicrobials' (WHO, 2000c). AMR forces health systems to undertake more expensive diagnostics, treatment and therapy, which might include (if available) more costly antibiotics. AMR-related expenses further harm many low-income countries' ability to deal with pneumonia.

Diarrhoeal diseases are also major killers in the low-income world, claiming over 2.2 million deaths annually (WHO, 1999b) and AMR has joined forces with these diseases to create a public health nightmare. Shigella, cholera and typhoid have all developed resistance to first-line antibiotic treatments. Shigella is resistant to almost every available antibiotic (WHO, 2000c); efforts to contain cholera are complicated by drug-resistant cholera strains (WHO, 2001a); and multi-drug resistant typhoid has become a major global problem (WHO, 1997a, 2000c).

Malaria kills over one million persons a year and approximately 300–400 million new cases of malaria occur annually (WHO, 2000c). Such mortality and morbidity happen mainly in low-income countries (WHO, 1998f). Malaria is a major cause of death in 92 countries, in 80 per cent of which malaria is resistant to the first-line treatment drug, chloroquine (WHO, 1999b, 2000c). Resistance to second- and third-line drugs continues to develop, threatening to leave malaria control without any effective antimalarial drugs (WHO, 2000c).

Drug-resistant tuberculosis (TB) has also emerged as a global public health concern (WHO, 2000d). TB kills approximately 1.5 million people annually (WHO, 1999b) and drug-resistant TB is adding to this death toll. Global levels of drug-resistant TB are hard to estimate because of poor surveillance and reporting, but WHO believes multi-drug resistant TB (MDRTB) accounts for 1–2 per cent (and growing) of current worldwide TB cases of approximately 16 million (WHO, 2000c). In certain regions, such

Table 9.1   Main causes of AMR development

Cause	Explanation
Process of natural selection	Microbes adapt to pressures placed on them by the environment, including competition from other microbes, human immune systems and antimicrobial drugs. Antimicrobials accelerate the process of natural selection by eliminating susceptible microbes and leaving resistant ones behind to thrive
Inappropriate use of antimicrobial drugs	Inappropriate use involves using (1) inadequate doses or duration of the correct drug, or (2) the wrong drug. An example is the utilization of anti-TB drugs in doses and durations that differ from international guidelines, such as directly observed therapy – short course (DOTS)
Overuse of antimicrobial drugs	Overuse involves using antimicrobials to treat diseases against which the antimicrobials have no effect. An example is the prescription of antibiotics to treat acute viral respiratory infections
Under-use of antimicrobial drugs	Under-use occurs when an effective drug is not used appropriately because of poverty or lack of affordable access to the drug. Poverty often means that malaria drugs are used improperly, and lack of affordable access exacerbates MDRTB's spread
Use of poor quality or counterfeit drugs	The global prevalence of poor quality or counterfeit drugs is enormous. Such drugs are not effective in eliminating infections and spur AMR development
Inadequate surveillance of resistance	Surveillance of resistance trends is critical to public health efforts to address the AMR problem, but national and international surveillance systems for resistant organisms are often inadequate or nonexistent
Inadequate diagnostic techniques	Proper antimicrobial use is dependent on a correct diagnosis of the disease-causing organism, but cost-effective diagnostic tools and technologies are often not available to assist the rational use of antimicrobials. This diagnostic problem often leads to antibiotics being prescribed for viral infections
Lack of new classes of antimicrobials	Public health experts perceive the need for new antimicrobials as the old ones lose effectiveness; however, pharmaceutical research and development on new drugs for malaria, TB, and other diseases plagued by resistance has been inadequate
Lack of vaccines	Vaccines represent a superior way to prevent infectious diseases and AMR development, but effective vaccines for TB, HIV/AIDS, malaria, and other diseases developing resistance have not been developed
Failure of infection control in hospital settings	Nosocomial infections spread because the health-care institution's infection control procedures are antiquated or are not appropriately followed by medical staff
Use of antimicrobials in food animal production	Huge amounts of antimicrobials are used to promote the growth of food animals, including cattle, poultry and fish. Such use contributes to AMR development for organisms, such as Enteroccoci, that cause infections in humans

as Eastern Europe and the former Soviet Union, the MDRTB rate is as much as one in five TB patients (WHO, 1999b). MDRTB 'hot zones' spell trouble for global TB control efforts. WHO fears that these zones 'could propel a wave of tuberculosis that is difficult – even impossible – to cure using drugs' (WHO, 1999b, p. 46).

This growing problem is not limited to resistance in leading infectious diseases. WHO also reports growing AMR in the treatment of viral hepatitis, hospital-acquired infections (especially methicillin-resistant Staphylococcus aureus and vancomycin-resistant Enterococcus), leishmaniasis and gonorrhoea (WHO, 2000c). WHO also fears that resistance will develop in helminthic diseases prevalent in the developing world (WHO, 2000c).

The diversity and complexity of the factors behind AMR emergence darkens these depressing trends. AMR arises from an array of causes that are difficult to catalogue, let alone analyse. Table 9.1 summarizes some of the most frequently mentioned causes of AMR development.

9.2 AMR as a global governance problem

The national and international public health actions plans on AMR stress three objectives: (a) improving surveillance of resistant organisms; (b) increasing the rational use of antimicrobials; and (c) strengthening basic science research in order to stimulate the development of new antimicrobials. These objectives form the 'grand strategy' for the public health response to AMR and constitute the framework for the global health governance (GHG) now developing on this issue. Surveillance is at the heart of infectious disease control, so the emphasis on surveillance in connection with AMR comes as no surprise. Public health experts need to know where the problem is, and the scale of resistance, in order to craft proper intervention strategies. More rational use of antimicrobials seeks to control and reduce inappropriate use of antimicrobials that takes place in human medicine and food animal production. Increasing rational use also aims to lengthen the effectiveness of existing antimicrobials before they become useless. Because the pipeline of new antimicrobial products has been fairly dry in recent decades, the global AMR crisis forces public health experts to confront the difficult research and development task of creating new, effective and safe drugs and vaccines.

The scale and seriousness of the AMR problem suggests that it requires radical and far-reaching responses from national governments, international organizations and nongovernmental organizations (NGOs). Not surprisingly, public health and other government officials have been formulating responses to the AMR crisis. WHO has been crafting a Global Strategy Against Antimicrobial Resistance (WHO, 2001b) and addressing the threat to human health posed by the use of antimicrobials in food animal production (WHO, 1997b, 2000e). The European Union (EU) banned the use in animal feed of four antibiotics that are important to human health (EU, 1997, 1998), a ban that a pharmaceutical company challenged before the European Court of Justice (ECJ, 1999). In 2000, the US Congress passed legislation that included provisions on fighting AMR (Public Health Improvement Act, 2000). In collaboration with other federal agencies, the US Centers for Disease Control and Prevention issued a Public Health Action Plan to Combat Antimicrobial Resistance (CDC, 2000); and legislation has been introduced in Congress to fund aspects of this federal Action Plan (Reuters, 2001). The US Food and Drug Administration has (a) proposed new labelling requirements for antibacterial drug products to warn of AMR (FDA, 2000a); and (b) reviewed its regulatory responsibility to ensure that the use of antimicrobials in food production does not harm public health (FDA, 2000b). At the civil society level, NGOs have formed, such as the Alliance for the Prudent Use of Antibiotics, to address the AMR problem and advocate solutions (APUA, 2001).

While ongoing AMR governance activities can be discerned, the approach being taken at the international and national levels may be inadequate to handle the AMR crisis. WHO's AMR work follows its historical penchant for creating scientific and technical standards and making non-binding recommendations that national governments should follow (WHO, 2001c). AMR does not feature in WHO's revision of the International Health Regulations (IHR), which are binding rules of international law. Under current proposals, the revised IHR would require WHO Member States to report public health risks of 'urgent' international importance (WHO, 2000f, 2001d). This might be defined in such a way that outbreaks of drug-resistant diseases would be caught. Improving surveillance for drug-resistant pathogens is not, however, the purpose of the IHR revision process. Further, historical compliance with IHR reporting requirements has been dismal (Fidler, 1999). Hoping that the revised IHR alone will help GHG on AMR would thus not be advisable.

Reliance on WHO non-binding recommendations to national governments as the primary means of global AMR governance is questionable. WHO reports that one of the reasons for MDRTB's development is that half the world's countries have failed to follow WHO's recommendations about the use of DOTS to control TB (WHO, 1999b). Of the half that have adopted DOTS, one-third have failed to make the treatment available on a countrywide basis (WHO, 1999b). Compliance with other WHO recommendations in the infectious disease area, such as HIV/AIDS education in schools, is equally disappointing (WHO, 1999b). Effective international action depends, in large measure, on the capacity or willingness of national governments to implement the required policies. The weak public health capacity in many low- and middle-income countries complicates international efforts to encourage and assist governments to adopt WHO recommendations, guidelines and advice on AMR governance.

Furthermore, national action in a few countries will not provide robust GHG on AMR. Because drug-resistant pathogens move about the planet (WHO, 1999b), national governance in a few countries will have only limited impact on the global nature of the problem (Fidler, 1998). Improved surveillance and rational drug use in the TB context in the US would not, for example, curtail MDRTB's spread from resistance 'hot spots' in other parts of the world. The EU's ban on the use of four antibiotics will have reduced effectiveness if food producers elsewhere continue to use antimicrobials for growth promotion. National AMR reforms thus remain vulnerable because of the global context of the problem. Successful national AMR strategies are dependent on support from robust international efforts in order to contribute to curbing AMR.

The interdependence of national and international governance required to deal with AMR is not novel, in either public health or other areas of international relations. The IHR failed in large part because WHO member states refused, or were unable, to comply with them at the national level. In the context of tobacco control, WHO has moved away from its traditional recommendatory and informatory role to propose a Framework Convention on Tobacco Control (FCTC), which will contain rules that bind states under international law and create a regime specifically for reducing tobacco consumption (WHO, 2001e). As discussed by Collin (in this volume), the hope is that the FCTC will provide a stronger mechanism through which to improve national tobacco control policies.

The interdependence of national and international governance can also be observed in the context of global environmental issues. Regulation of global environmental pollution and degradation has not been left to non-binding recommendations by international organizations or the uncoordinated actions of individual states. International environmental law (IEL) has developed to provide procedures, institutions and substantive rules for global governance against environmental threats. Indeed, WHO has drawn from IEL in formulating the FCTC as the cornerstone of GHG on tobacco.

The remainder of this chapter analyses whether IEL provides any templates for GHG on AMR. This analysis contemplates a GHG approach radically different from what is currently happening or being discussed in global anti-AMR activities.

9.3 Global governance and AMR: a global public goods approach

At first glance, thinking about AMR in connection with global governance on the environment might seem odd. IEL exists to protect natural resources, such as rainforest or the ozone layer, from anthropogenic harm. Antimicrobials are not natural resources but are creations of human ingenuity. The link between AMR and governance on global environmental threats appears when we conceive of anti-AMR efforts as seeking to preserve an anthropogenic resource – antimicrobial effectiveness (AME). Economists see in the AMR problem the classic 'tragedy of the commons' scenario. People misuse a common resource – an antimicrobial – to gain short-term benefits at the expense of the long-term destruction of the resource (*The Economist*, 2001). The antimicrobial 'commons' is, however, global in scope, meaning that efforts to preserve AME in one country will be inadequate from a public health perspective. International co-operation is needed to address the global destruction of the antimicrobial 'commons'.

Seeing AME as a common resource fits well with analysing AMR as a consequence of underproduction of a 'global public good'. The global public good in this case is AME; but AMR development across many infectious diseases indicates that states, international organizations and non-state actors underproduce AME. In the typology of public goods, AME is an impure public good or a common pool resource because (a) access to the good is (theoretically) available to all; but (b) consumption is rivalrous as the good can be depleted (Kaul *et al.*, 1999).

Problems that low-income countries have with access to affordable antimicrobials do not mean that AME is not a public good. As Section 1 suggests, access problems indicate that the public good of AME is underproduced. Patent protection for new pharmaceutical products complicates a public goods analysis of antimicrobials, because the patent limits access and makes drugs under patent protection look more like private goods (that is drugs for which access is excludable and consumption is rivalrous). The thrust of campaigns by international organizations and NGOs to increase access to drugs for infectious diseases, such as HIV/AIDS, TB and malaria, is to ensure that access to such essential drugs is open to all countries and that such drugs are public not private goods.

As public goods literature demonstrates, public goods are often underproduced because of 'market failure' (Kaul *et al.*, 1999). No single consumer of an accessible but degradable resource has sufficient incentive to limit consumption for the resource's long-term preservation. In other words, the market fails to provide adequate incentives for resource conservation, which threatens to destroy the resource unless government intervenes to regulate consumption. This dynamic arises in the global use of antimicrobials. Open and essentially unregulated access to antimicrobials produces abuse of the resource that leads to the depletion of its public health effectiveness. To conserve AME requires government intervention at the national level and co-operation among states and non-state actors globally.

Because international co-operation is critical to the supply of global public goods, experts stress the importance of 'international regimes' to the production of global public goods (Kaul *et al.*, 1999). IEL, for example, is composed of different regimes on various international environmental problems, illustrating the importance of constructing international law to deal with the global depletion of common resources. WHO officials stress the FCTC's importance as an international legal regime to support production of the global public good of tobacco control (Taylor and Bettcher, 2000). Infectious disease control is also a global public good, and presumably the revised IHR will constitute the international legal regime to improve production of this global public good. The common theme in these examples is the use of international law by states and international organizations to produce global public goods.

To date, the international AMR regime mainly involves non-binding recommendations and guidelines from WHO. If international law is important to the production of global tobacco and infectious disease control, should WHO rely on non-binding recommendations in its global AMR strategy? Conceptualizing AME as a global public good raises the question whether the current WHO approach on AMR is sufficient. Because it focuses on the preservation of resources, IEL offers a radically different path for GHG on AMR that should be examined when thinking about what states, international organizations and non-state actors should do to fight the global spread of AMR.

9.4 Lessons for GHG on AMR from international environmental law

AMR and the different kinds of international environmental problems

International environmental problems generally fall into one of three categories: (a) a country engages in the unsustainable exploitation of a national resource, such as a rainforest, that other states have an interest in preserving; (b) domestic activities within a country, such as power generation, release pollutants that cross borders and cause damage in other states; and (c) domestic activities taking place in many countries threaten environmental resources used by all humankind, such as the ozone layer or the global climate system (Fidler, 2001). AMR exhibits characteristics from all three international environmental problems and makes international law created for these problems relevant for thinking about AMR control.

AMR can be analogized to the unsustainable exploitation of a national resource, such as rainforest. A rainforest, such as the Amazon, extends beyond the borders of a single state; but a state, such as Brazil, determines the exploitation of the rainforest within its boundaries. Similarly, an antimicrobial drug, such as chloroquine, is a resource that many states consume; but the consumption in any one state is determined by that state's exercise of sovereignty. The unsustainable exploitation of either rainforest or antimicrobials nationally becomes an international concern because the resource's depletion adversely affects environmental or public health interests far beyond that state. International law developed to address the unsustainable exploitation of national environmental resources may thus be relevant to thinking about GHG for AMR.

The AMR problem also resonates with environmental concerns about transboundary pollution because resistant microbes travel across borders. In the transboundary pollution context, domestic activities, such as operating a factory, harm the environment and public health inside other states because domestically produced pollution crosses borders. Similarly, the domestic use of antimicrobials may cause public health damage inside other states because antimicrobial misuse can generate transboundary traffic in resistant pathogens. The resistant microbes become transboundary pollutants. Thus, international law on transboundary air and water pollution might be helpful in thinking about GHG for AMR.

Finally, AME might be likened to a global resource such as the ozone layer or the global climate system. For example, the ozone layer is a global resource, the use and benefits of which are available to all countries. Ozone-layer depletion occurs because many countries emit pollutants from domestic activities that damage the layer. Similarly, misuse of antimicrobials in multiple countries damages AME globally. International law on ozone depletion and the protection of other global environmental resources, such as the global climate, might be relevant to GHG on AMR.

International law on unsustainable exploitation of national environmental resources as a template for GHG on AMR

Under traditional international law, a state has sovereignty over its territory and population (Brownlie, 1998); and other states should not intervene in matters that are within the domestic jurisdiction of another state (UN, 1945). In the context of natural resources, states emphasized this authority through international legal rules, such as permanent sovereignty over natural resources (UN, 1962). The existence of these rules meant that international efforts to address the unsustainable exploitation of national environmental resources took place through treaties because customary international law (CIL) provided no foundation for action. Similarly, how a state utilizes antimicrobials would, under CIL principles, be a matter within the state's domestic jurisdiction in which other states have no legal right to intervene. CIL provides no basis for states to challenge a country whose domestic public health policies (or lack thereof) produce, or threaten to produce, AMR.

The international environmental treaties concerned with unsustainable exploitation of national resources, including the UN Convention on Desertification (1994) and the UN Convention on Biodiversity (1992), essentially consist of a deal between high- and low-income countries. The bargain calls for substantive commitments by: (a) low-income countries to take action to reduce the unsustainable development of the resource; and (b) high-income countries to provide the low-income countries with financial and technical assistance to help them preserve the resource in question (Fidler, 2001). The treaties also establish institutional machinery – conference of the parties, secretariat and committees – to oversee the implementation of the bargain.

Antimicrobial misuse is not confined to low-income countries. High-income countries experience antimicrobial abuse in both human medicine and food and animal production (Okeke and Edelman, 2001), making duties to protect AME relevant to high-income countries. Nevertheless, the template from IEL on unsustainable exploitation of national resources suggests that the high-income/low-income country bargain struck in environmental treaties is relevant for thinking about GHG on AMR. Low-income countries need financial and technical assistance to improve resistance surveillance and rational use of antimicrobials, and such help will only be forthcoming from the developed world. Further, the obligations used in the international environmental treaties, such as giving priority to combating desertification and destruction of biological diversity and establishing national action programmes, would be the kind of duties GHG on AMR could usefully adopt.

International law on transboundary pollution as a template for a GHG on AMR

Transboundary pollution usually exhibits two important factual features (Fidler, 2001). First, the polluting activity occurs within the territory of a sovereign state. Second, the polluting activity creates environmental and/or public health problems in other countries, making the pollution an international concern. These aspects converge to produce the proper policy response: the state in which the polluting activity occurs has to alter its policies to reduce or eliminate the threats the pollution creates.

The transboundary pollution model is useful for thinking about GHG on AMR. Domestic misuse of antimicrobials produces resistant pathogens that migrate across borders to cause public health harm in other states. The transboundary flow of resistant organisms between high- and low-income countries runs in both directions (Okeke and Edelman, 2001). Antimicrobial misuse produces transboundary 'pollution' in the form of drug-resistant microbes. To stem the flow of transboundary microbial pollution, the state in which the antimicrobial abuse occurs has to alter its policies to reduce or eliminate the public health threats the microbial pollution creates.

Rules on transboundary pollution are found in both CIL and treaty law. Under CIL, Birnie and Boyle (1992, p. 89) argued that '[i]t is beyond serious argument that states are required by international law to take adequate steps to control and regulate sources of serious global environmental pollution or transboundary harm within their territory or subject to their jurisdiction'. This CIL principle ostensibly imposes a duty on all states to prevent, reduce and control environmental harm that their domestic activities might cause to other states.

Whether this CIL rule exists 'beyond serious argument' is, however, questionable. Bodansky (1995, pp. 110–11) has argued, for example, that:

> transboundary pollution seems much more the rule than the exception in interstate relations. Pollutants continuously travel across most international borders through the air and by rivers and ocean currents. In a few cases, states have undertaken efforts to reduce these pollution flows – generally, through treaties. Leaving aside the question of whether these treaties can create or are evidence of a customary norm, they apply to relations among a small fraction of the 180-plus countries of the world, and presumably cover only a small part of the total flow of transboundary pollution. As Schachter [1991] concludes, 'To say that a state has no right to injure the environment of another seems quixotic in the face of the great variety of transborder environmental harms that occur every day.'

Based on the purported CIL rule obliging states to prevent, reduce and control environmental harm that their domestic activities cause other states, we could perhaps argue that CIL likewise imposes a duty on states to prevent, reduce and control the antimicrobial misuse within their jurisdictions because such misuse can cause harm in other states. However, this argument is not persuasive. First, global efforts on AMR are so recent that a sufficient body of state practice has not yet developed to support a CIL obligation on AMR. Even the growing concern and regulatory effort surrounding antimicrobial use in food animal production does not yet amount to the general and consistent state practice required for the formation of a CIL rule. Second, the tension between the CIL rule on preventing transboundary pollution and the day-to-day reality of such pollution suggests that CIL is not a solid foundation on which to base GHG on AMR. Even if it was plausible that CIL imposes a duty on states to prevent, reduce and control antimicrobial misuse, the reality – in both high- and low-income countries – is widespread misuse.

The problems encountered with CIL lead AMR strategy to treaty law, just as states have turned to treaties to advance global governance on transboundary pollution. It is beyond the scope of this chapter to discuss comprehensively the treaty law on transboundary pollution, so this analysis focuses on the main features of these regimes. The general objective in these treaties is to prevent, reduce and control transboundary pollution. This objective echoes the purported CIL rule examined above. The treaties attempt, however, to give this obligation meaning through procedural and substantive duties.

Treaties on transboundary pollution typically contain procedural duties that require states' parties to co-operate, to exchange information, to notify other states parties of certain events and to consult with respect to specific activities. Institutional machinery (for example secretariat, conference of the parties, scientific committees) created in the treaty to oversee co-operation supports these procedural duties. In transboundary pollution treaties negotiated under the auspices of the UN Economic Commission for Europe (UNECE), the procedural duties and institutional machinery are set up in 'framework' conventions, namely the 1979 Geneva Convention on Long-Range Transboundary Air Pollution (LRTAP) (UNECE, 1979) and the 1992 Convention on the Protection and Use of Transboundary Watercourses and International Lakes (Water Treaty) (UNECE, 1992). These framework conventions establish ongoing diplomatic, legal and scientific processes that lead the states parties to take more substantive action against transboundary air and water pollution.

Substantive duties that require states' parties to take specific actions to prevent, reduce or control transboundary pollution appear in protocols to the framework conventions. In the LRTAP regime, states' parties have negotiated, to date, protocols to reduce by specific amounts the release of nitrogen oxide, sulphur, volatile organic compounds, heavy metals and persistent organic pollutants (UNECE, 2001). The states' parties have also negotiated a 'multi-effects, multi-pollutant' protocol that seeks to cut emissions of sulphur, nitrogen oxides, volatile organic compounds and ammonia (UNECE, 2001). In the water pollution context, states parties to the Water Treaty negotiated the Protocol on Water and Health (UNECE, 2001). The framework-protocol approach to transboundary pollution attempts to bring states' parties progressively closer to the actual reduction of crossborder pollution.

This sketch of treaty law on transboundary pollution raises the question of whether the framework-protocol approach might be promising for GHG on AMR. The framework convention could address the general AMR problem, while protocols could deal with AMR control in specific diseases. Before I address the framework-protocol strategy further, it is necessary to look at international law on the protection of common global resources to see whether this law is also relevant to constructing a global governance strategy on AMR.

International law on protecting global resources as a template for GHG on AMR

Degradation of global environmental resources typically exhibits three important features (Fidler, 2001). First, the environmental resource in question does not fall within any one state's jurisdiction, nor is it confined to the jurisdictional reach of a few states. The resource is a commonly available and accessible resource – the planet's oceans, ozone layer and climate. Second, the resource is being degraded by activities taking place within sovereign states that are subject only to the jurisdiction of the state in which they occur. Third, the degradation of the resource is multilateral in nature because pollution from many states produces the damage. This fact implies that pollution reduction must also be multilateral to address the degradation effectively.

At first glance, it may seem odd to think of AME as a global resource in the same way we think of the ozone layer. The ozone layer is a natural resource, while antimicrobials are manmade. In addition, patent rights on antimicrobials mean that, at least during the patent's life, antimicrobials are protected as private property. Despite these differences, it is still helpful to think of AME as a global resource because antimicrobials become globally available both before and after patent protection expires. Given the global nature of microbial traffic, the degradation of AME in one part of the world threatens public health in other parts, in the same way that degradation of the ozone layer by pollution from one country threatens populations in other countries. For this reason, the analogy between natural global resources and AME is useful.

Protecting common environmental resources has forced states and international organizations to develop and implement new international legal principles and approaches. The traditional CIL rule prohibiting transboundary harm from pollution does not apply to degradation of global environmental resources because the resource being harmed does not fall within the jurisdiction of any state. The international legal principles of sovereignty and non-intervention also mean that the domestic activities causing harm to a global resource are within the sovereign state's exclusive control. Further complicating matters is the fact that, under international law, global resources are open for use by all states (for example fishing on the high seas). Traditional principles of international law do not provide sufficient deterrents or incentives for states to reduce their degradation of a common resource. The result is the famous 'tragedy of the commons' on a global scale.

The global governance challenge in connection with the degradation of global environmental resources has been threefold (Fidler, 2001). First, duties needed to be created to get states to address the degradation of environmental resources beyond their jurisdictional control. Second, ways needed to be found to make the duty to address degradation of global environmental resources effective from a scientific perspective. Third, the proper incentives had to be crafted to encourage states to make the economic sacrifices necessary to protect the resource.

International environmental lawyers argue that CIL contains a duty requiring states to prevent, reduce and control pollution of areas beyond national jurisdiction (Birnie and Boyle, 1992). This purported CIL rule is expressed in Principle 21 of the Stockholm Declaration (1972) and can also be found in other international legal documents. But this CIL norm suffers the same problems that the identical CIL rule applicable in the transboundary pollution does – the norm does not reflect state practice and is too general to provide the specific pollution-reduction actions needed to mitigate the degradation.

Attention turns again to treaty law, and we see states and international organizations using the 'framework-protocol' approach to tackle depletion of the ozone layer and global warming – two of the most serious forms of degradation of global environmental resources. As in the transboundary pollution context, the framework conventions for ozone depletion (Vienna Convention, 1985) and climate change (Climate Change Convention, 1992) contain (a) procedural duties of co-operation, information exchange, notification and so on and (b) institutional machinery for pushing the regime toward more substantive duties. Protocols to the framework conventions contain the substantive duties that require states to reduce emissions of pollutants by specific amounts by certain dates (Montreal Protocol, 1987; Kyoto Protocol, 1997).

One problem encountered with the framework-protocol approach in connection with ozone depletion and global warming has been the need to offer special incentives to low-income countries to encourage them to participate. The ozone and climate change regimes offer two kinds of incentives: (a) financial and technical assistance to low-income countries to help them implement and comply with the treaty rules; and (b) differential duties under which low-income countries' treaty obligations are less onerous than those of high-income countries. In the climate change regime, for example, low-income countries are currently bound by no more than general procedural and institutional obligations in the framework convention (Climate Change Convention, 1992), while the Kyoto Protocol (1997) imposes duties to reduce greenhouse gas emissions only on high-income countries.

The international law on the degradation of global environmental resources suggests that global governance for AMR needs: (a) procedural duties commonly found in the framework conventions; (b) institutional machinery typically established in the framework conventions; (c) an ongoing scientific and legal process that develops progressively more specific duties on the prevention, reduction and control of AMR; (d) financial and technical assistance to low-income countries; and (e) less onerous burdens for low-income countries. The analysis of the international law on transboundary pollution supports the need to include items a–c in any global strategy against AMR. The lack of public health preparedness in the low-income world on AMR also makes items d and e relevant for a global governance strategy. However, whether less onerous duties for low-income countries are appropriate may be challenged on scientific and public health grounds because such leniency would provide opportunities for AMR to develop and spread.

How well do the templates from IEL fit the AMR problem?

The preceding analysis assumed that AMR could be analogized to the three general kinds of international environmental problems. But is there sufficient fit between AMR and unsustainable exploitation of national resources, transboundary pollution and degradation of global environmental resources to see much value in the international legal regimes addressing these international environmental concerns for GHG on AMR?

Again, public health's 'grand strategy' for dealing with AMR involves (a) improved surveillance, (b) more rational use of antimicrobials, and (c) development of new antimicrobial drugs. Surveillance is the collection and analysis of epidemiological data. For this data to be useful, it must be shared locally, nationally and internationally. The procedural duties, institutional machinery and ongoing processes found in IEL would support the goal of improved surveillance. Thus, the legal infrastructure suggested by IEL could assist the public health goal of improved surveillance of resistant pathogens. The weakness or non-existence of public health infrastructure in low-income countries will also require technical and financial assistance to such countries if surveillance is to be improved internationally. As in the environmental context, the global governance strategy on AMR has to include financial and technical assistance to low-income countries as part of bringing them into the regime.

The objective of more rational use could also be supported by the legal architecture implied by IEL on unsustainable exploitation of national resources, transboundary pollution and degradation of global environmental resources. Procedural duties, institutional mechanisms and an ongoing diplomatic process could raise the profile of how countries deal with rational-use problems. The legal framework could also support substantive duties, such as the requirement that states parties mount AMR education campaigns. The relative success and failure of these campaigns could be analysed through the ongoing scientific and legal process. There may also be the potential to develop other substantive duties relating to the use of antimicrobials in food and animal production. The AMR regime could require, for example, that WHO's recommendation on the termination of antimicrobial use in animals that threaten resistance in human pathogens (WHO, 1997b) be implemented through national law by states parties. As with surveillance, low-income countries would need technical and financial assistance to fulfil such substantive legal duties.

To encourage the development of new antimicrobials, a global AMR regime could support pursuit of this goal in a number of ways. First, the procedural duties, institutional mechanisms and ongoing diplomatic process could provide a supportive environment to (a) assist public–private partnerships on the development of new antimicrobials, and (b) address ways to create sufficient economic incentives for pharmaceutical companies to develop new drugs without adversely affecting access in the developing world. Second, the AMR regime could provide ways to deal with the interdependence of new drug development and more rational use. The lack of public health infrastructure in low-income countries to control rationally antimicrobial use erodes the economic incentives for pharmaceutical companies to develop new drugs. Pouring huge amounts of money into the development of antimicrobials that will be used globally in uncontrolled ways is not a prudent public health approach to preserving AME.

A treaty-based AMR regime could also provide opportunities and leverage for increased global civil society activities. International environmental treaties have stimulated environmental NGO activities, and WHO's efforts on the FCTC include using the negotiations for this treaty to catalyse global civil society action against tobacco consumption and the behaviour of tobacco multinationals (see Chapter 5). The same dynamic could be created and harnessed for GHG on AMR.

This brief analysis suggests that the international legal templates offered by IEL logically apply to the scientific and public health objectives that effective GHG on AMR would have to pursue.

9.5 Conclusion

This chapter's analysis of IEL as a template for GHG on AMR points in a direction radically different from anything currently being discussed or considered. In fact, many in the public health community are determined to expand access to antimicrobials in low-income countries for HIV/AIDS, TB, malaria and other infectious diseases. This strategy represents potential danger for GHG on AMR. WHO and leading NGOs, such as Médecins Sans Frontières (MSF), for example, support greater access to antimicrobials as a human right. Increasing access to antimicrobials dominates WHO's recommendations under its 'widely and wisely' approach to AMR control. Of the eight specific recommendations in WHO's approach, four relate directly to increasing access to antimicrobials for people in low-income countries (WHO, 2000c). While there are clear and urgent needs for improved access, the human rights rhetoric may obscure the scientific and public health complexity that the use of antimicrobials involves. In addition, campaigns for access proceed even though the public health infrastructure in many countries to conduct surveillance for resistant organisms and to ensure rational use of antimicrobials is weak or does not exist at all. While creative initiatives to increase access and rational use are underway, as in the WHO/MSF/Harvard Medical School effort to improve access to and rational use of second-line drugs for treating MDRTB (Gupta *et al.*, 2001), the sustainability of such activities remains uncertain.

The emphasis on access is not, however, to blame for the lack of progress on improved surveillance and rational use at the national governmental level. Rich countries, such as the US, also face enormous public health, financial and regulatory challenges in improving national AMR governance. The similar tasks confronting poorer countries grow in magnitude as the available human, governmental and financial resources decrease. WHO's experience with obligatory infectious disease surveillance in the

IHR provides no comfort for those wishing to mandate surveillance on drug-resistant organisms through international law. International legal obligations for AMR control modelled on international environmental agreements would face the difficulties environmental treaties have suffered, including inadequate national implementation and compliance with treaty rules. In short, the creation of a global governance regime on AMR based more firmly in international law will not provide a 'magic bullet' against the global spread of AMR.

AMR has been only recently recognized as a challenge for GHG. The scale, scope and adverse consequences of AMR development across a range of infectious diseases signal a public health crisis of global proportions. Controlling AMR is a global governance challenge because such efforts seek to reduce the production of the global public bad of AMR and to increase the supply of the global public good of AME. The production of global public goods requires international co-operation, and very often involves the use of international law to create the procedural, substantive and institutional frameworks needed to mobilize political will and resources. The existing global governance strategy on AMR follows WHO's traditional approach of promulgating non-binding advice and encouraging countries to adopt its recommendations. Beyond arguing that access to antimicrobials is a human right, WHO has not, to the current author's knowledge, considered grounding its global governance strategy on AMR in international law, the way in which it has for infectious disease control through the IHR and for tobacco control through the FCTC. The magnitude and nature of the global AMR crisis suggests, however, that GHG on this issue cannot be adequately addressed alone through good-faith efforts at persuasion.

The evolutionary powers of pathogenic microbes are immense. The capabilities of people and governments to create public health problems that augment microbial evolution are also immense. While more robust GHG on AMR is necessary, it will not rescue humanity from the evolutionary powers of nature, or the fallibility of humans and their institutions. We will never 'out-govern' pathogenic microbes. But we would not be smart in engaging in the AMR battle without acknowledging the seriousness of the problem and crafting a global governance strategy that is informed by existing precedents, and designed to undertake the public health equivalent of the labours of Sisyphus.

10
Assessing the Health Policy Implications of WTO Trade and Investment Agreements

Meri Koivusalo

10.1 Introduction

The multilateral trade agreements (MTAs) under the responsibility of the World Trade Organization (WTO) provide a binding legal framework for overseeing trade relations worldwide. However, MTAs concern a much broader array of policy measures than tariffs and customs, and as a consequence have much further reaching policy implications than often is recognized. In particular, they can be seen as the main legal structure for consolidating economic integration across countries, and ultimately promoting processes of economic globalization. As such, as Held *et al.* (1999) point out, the global regulation of trade by bodies such as the WTO implies a significant renegotiation of the Westphalian notion of state sovereignty (Held *et al.*, 1999).

In comparison with other international agreements, the MTAs are strengthened by the WTO's authority to impose trade sanctions and other forms of legal enforcement. For example, unlike global environmental agreements, International Labour Organization (ILO) conventions, or human rights commitments under the United Nations (UN) system, the WTO agreements are backed by stronger mechanisms to ensure compliance and dispute settlement. In practice, this means they have greater influence at the global and national levels than international agreements that are more declarational. In terms of global governance and rule-setting, the role of MTAs are central.

The nature and extent of conflict between the aims of the MTAs and other international agreements has been debated in relation to human rights and the environment, for example Biodiversity Convention and Biosafety Protocol (Final Report, 2000; Phillips and Kerr, 2000; UN, 2001; UN Commission on Human Rights, 2001). Attention has also been drawn to the more systematic implications of WTO agreements on public health policies (Bettcher *et al.*, 2000; Correa, 2000b; Koivusalo, 1999). For the most part, trade-related health issues have focused on risks from infectious diseases, migration of health professionals, harmonization of public health and safety regulations across states, or the health benefits of economic growth from trade.

This chapter argues that health matters and trade should been seen as a systematic global policy issue rather than a matter of particular nation states or developing countries. It is further argued that the interaction between global and local influences relate more to the level of health protection and nature of health policies defined by the state than to issues of national sovereignty. The real dilemma in the context of trade and investment agreements is thus not in the interface between national and global levels, but between public interests and public health policies and commercial and private sector profit interests reflected often in national trade and export emphases. While health policy measures and practices have remained mostly a national responsibility governed by national level policies, trade and commercial law have become transnational.

The first section of this chapter briefly describes the WTO and the key health issues arising from its various MTAs. While health figures in a relatively minor way in the MTAs, their underlying principles, rules and procedures have far reaching implications for public health. This is followed by a discussion of Finland as a case-study of how national health policies of one country can be shaped by the WTO. The chapter concludes by exploring the challenges of analysing the health implications of WTO trade agreements.

10.2   The health policy implications of the World Trade Organization

The WTO was created in 1995 as the successor to the General Agreement on Tariffs and Trade (GATT) established in 1946. The decision to establish the WTO was taken at the Marrakesh Ministerial Conference in 1994, which followed the completion of the Uruguay Round of GATT negotiations. However, the WTO is more than simply a continuation of GATT, for unlike its predescessor, the WTO has an institutional structure with full and permanent commitments. The essential functions of the WTO include administering and implementing the multilateral and plurilateral trade agreements which together make up the WTO, acting as a forum for multilateral trade negotiations, seeking to resolve trade disputes, overseeing national trade policies, and co-operating with other international institutions involved in global economic policymaking. The structure of the WTO is dominated by its highest authority, the Ministerial Conference, which is composed of representatives from all WTO member states. This body is required to meet at least once every two years. The WTO is, in principle, an intergovernmental organization as only member states may take matters to the dispute settlement process. Decisionmaking is officially based on the one country one vote system, although most decisions are taken by consensus.

There are a wide range of implications for health policy raised by the WTO and its MTAs. Most fundamentally, perhaps, is the underlying philosophy of the organization based on the assumptions that open markets, non-discrimination and competition in international trade are conducive to the welfare of all countries. Embedded in this assumption is the residual nature of social policy and, in particular, the redistribution of generated wealth as part of national social policies. It is thus concluded that trade leads to economic growth which, in turn, leads to poverty reduction and better health. This has been reflected strongly in recent analyses of globalization and health (Dollar, 2001; Feachem, 2001). Challenging this perspective is evidence that globalization and liberalization of markets is more closely associated with increased social inequality (Cornia, 1999). In practice also linkages between international financial and trade institutions have become closer and more complex than ever (Ahn, 2000). This accompanies long established research showing that relative equity in the distribution of wealth within and between societies is an important determinant of health (Blane, 1996; Pearce, 1996; Tesh, 1988; Wilkinson, 1996). The challenges of regulating transnational corporations at the global level and the indirect implications of trade liberalization on the performance of the public sector (Pollock and Price, 2000) also raise difficult questions for health policy.

Given the *raison d'etre* of the WTO, its stipulations are based foremost on the promotion of trade and the interests of member states in relation to trade matters. What is particularly evident is the role of corporate lobbying in WTO meetings (Marceau and Pedersen, 1999; Scholte *et al.*, 1999), such as in negotiations on the Agreement on Trade-Related Aspects of Intellectual Property Rights (TRIPS) (Drahos, 1995, 1997). Sklair (2001) defines such interests as the 'transnational capitalist class', which seek to minimize national restrictions on their wealth generating activities. While WTO agreements contain clauses concerned with public health, these are very limited and only relate to selected aspects of health policy. WTO stipulations complicate equity-oriented health policies and service delivery through the primacy concern of equity between corporate service providers seeking to trade across countries in comparison to equity considerations related to access and costs of service provision (WTO, 1994c). Thus, while WTO agreements contain public health clauses, these cannot be seen as sufficient or only aspects of health policies in practice.

WTO Agreements permit public health and other regulatory measures, but such measures are mostly required not to restrict trade. While it would be prudent to propose that such measures should not unnecessarily restrict trade, the emphasis on limiting trade restrictions has direct health policy implications. This emphasis promotes less trade restrictive individualized measures and more residual policy choices at the expense of broader public policy approaches, thus influencing directly the quality and nature of health policies (Koivusalo, 1999).

As well as having very different starting points and ultimate goals, the rules governing trade raise specific concerns for the making of health policy. One issue concerns how 'discrimination' is defined in the MTAs. Whether it concerns trade of goods or services, the emphasis in judging discrimination is based on the 'likeness' of the products or outputs, with no consideration of the production processes or ownership. For example, all producers of like products are treated in a similar way whether or not they engage in potentially unhealthy production practices, such as the use of hormones or antibiotics in cattle raising. This is especially so where there are no known public health implications for the end product.

Another issue concerns the fundamental conflict of interest between many areas of health regulation and promotion, and the aims of the corporate sector to maximize their markets. In some cases, the products themselves have harmful or potentially harmful effects on health. The tobacco industry is a good example of an industry that manufactures and trades a product worldwide that harms human health. Similar concerns are raised in relation to the inappropriate marketing of alcohol, baby milk, unhealthy fast-food products and sugary drinks, notably to children. The aim of health promotion policies is to limit consumption of such products, which means a potential restriction to trade. Such policies do not fit squarely within WTO rules regarding allowable measures to address an immediate threat to public health. As such the undertaking of health promotional measures may become problematic within the trade-oriented context of the WTO.

The WTO dispute settlement process leads eventually to both the development and interpretation of international case law. In contrast to NAFTA only countries may take matters to dispute settlement, but it is a closed process dominated by trade interests and expertise. The dispute settlement process has no obligatory public health consultation and the expertise of panels is limited to senior trade figures. This means that any case law relating to the trade restrictiveness of public health policy measures will be adjudicated by senior trade officials and not by public health experts (WTO, 1994d). The analysis by Carlos Correa on WTO dispute settlement process and its implications for the scope of public policies that governments can undertake also clearly points out the problems of current practices (Correa, 2000a).

10.3 Agreement on Trade-Related aspects of Intellectual Property Rights

The specific treaties of the WTO are also worth considering in terms of their implications for health policy. The Agreement on TRIPS, signed in 1994, sets the ground rules for the protection of trademarks, copyrights and patents.

While the WTO is often considered an organization of deregulation, TRIPS is a good example of a treaty that creates greater regulation (WTO, 1994a). The implications of this agreement for public health are a growing subject of debate. One key criticism is the agreement's origins, which, critics argue, lie in the profit-seeking aims of multinational companies (MNCs) seeking to globally commercialize intellectual property rights, rather than being a product of a careful co-ordinated economic analysis (Drahos, 1995, 1997).

This is well illustrated by the protection of patents. A patent is, in principle, a reward for innovation. The basic function of a patent is to protect from other potential users the right of an individual or institution to benefit commercially or otherwise from an innovation for a given period of time. Pharmaceuticals under patent protection, in other words, can be sold at a higher price given the lack of competing products. TRIPS provides patent protection for 20 years, a level of protection reflected in the prices of most newer drugs on the current market. In developed countries pharmaceutical costs have been rising in spite of an emphasis on increasing competition for generic medicines, which are off patent. In developing countries pharmaceuticals listed in the WHO essential drugs list are mostly off patent, though the HIV/AIDS drugs are a major exception as most of them are newer innovations. While the relative share of public funds going to pharmaceuticals is high in least developed countries, representing up to 60 per cent of health budgets (WHO, 1998g), the costs of intellectual property rights are paid predominantly through higher pharmaceutical prices in developed countries.

According to the Organization for Economic Cooperation and Development (OECD), two-thirds of pharmaceutical products are purchased using public funds. In most OECD countries costs of pharmaceuticals have been rising steadily as a percentage of GDP since the 1970s (Jacobzone, 2000; Tarabusi and Vickery, 1998). While patents aim to reward research and development (R&D) costs, little is known about the distribution of costs of pharmaceutical R&D efforts in the corporate sector. The indicative high returns on investments in pharmaceutical research and biotech companies, as well as the higher share of marketing costs and shareholder returns in comparison to R&D costs in the 1990s, suggest some scope for reassessment (Public Citizen, 2001; Tarabusi and Vickery, 1998; UNDP, 2001). If governments wish to support private sector R&D, this may be done more explicitly through industrial policies and corporate support rather than through health budgets.

The problems of patenting and pharmaceutical pricing are global rather than specific to developing countries with high incidence of HIV/AIDS. However, given the high costs of new HIV/AIDS drugs, and the fact that a large majority of HIV/AIDS cases are in the developing world, patents have become a particular concern. This is especially so in middle-income countries where effective drug distribution is more possible. The dispute between

Brazil *versus* Roche is a good example. In 2001 the Brazilian government threatened compulsory licensing to lower the price of an HIV/AIDS drug produced by the pharmaceutical company Roche, the cost of which represented almost one-quarter of the HIV/AIDS programme budget (*Financial Times*, 2001a). There are attempts to interpret and apply TRIPS provisions to give greater attention to public health needs (Correa, 2000b; Velasquez and Boulet, 1997, 1999), and such efforts need to be continued (see Thomas chapter in this volume).

While the TRIPS is supposed to reward R&D efforts, there are increasing concerns over the problematic implications of TRIPS and patenting practices to R&D practices. The commercially driven incentive of patent protection has repercussions as the aim of R&D becomes driven more by markets than by health needs. One aspect of this has been the greater privatization of research through contract research organizations (CROs), which are less independent and cheaper than, for example, academically based institutions. In the US, 60 per cent of industry research grants go to CROs. Such contracts are closely tied to the needs of corporate clients (*Financial Times*, 2001b) that, as profit-seeking concerns, seek to respond to market demand rather than health need. As a result, R&D is heavily invested in treatments for the health concerns of the relatively wealthy (such as impotency, baldness, obesity), and far less for many life-threatening diseases with either low prevalence or limited commercial return.

The need by the pharmaceutical industry for patent protection and large profit margins is defended by the claims of an average R&D cost per drug of US$500 million. However, the high R&D costs are challenged by OECD figures showing that marketing costs and profits rose more than R&D costs during the 1990s (Jacobzone, 2000; Tarabusi and Vickery, 1998). The NGO, Public Citizen, challenges the US$500 million cost estimate, claiming that marketing and profits are driving up the costs of pharmaceuticals. It also points to the large contribution of publicly funded research in several important medicines (Public Citizen, 2001). Thus, the precise distribution of costs needs to be better understood in the pharmaceutical industry.

Another concern with respect to R&D is the accessibility of data and limitations of knowledge dissemination as the R&D sphere becomes increasingly privatized. It is known that industries also use patents and data protection clauses to block activities and defend interests, often acting against the broader public interest in research and knowledge accumulation. The TRIPS agreement obliges regulatory agencies to protect commercially valuable information when pharmaceuticals are approved for sale in countries. This can also be used to limit access to information, which could be of importance in, for example, assessment of the cost-effectiveness considerations or policy concerns. Concerns over access to data and legal challenges to regulatory agencies or health technology assessment measures have also been presented in the context of international trade agreements

(Hemminki *et al.*, 1999). It may be expected that as the fate of specific medicines and regulatory decisions on them become more important for the stock evaluations of pharmaceutical industry, pressures towards regulatory agencies will grow.

While most attention so far has been directed at the implications of patents under TRIPS, the protection of trademarks also raises public health concerns. Trademarks are an important part of the marketing worldwide of certain products, including global branding, promotion and sale. Public health policies may seek to limit the sale of products with harmful health effects through the regulating of marketing practices. A good example of this is control of marketing and promotion of alcohol and tobacco, for which a wide range of techniques based on trademark protection are used. The practice of 'brand stretching', where well-known brands are used to market other products (such as Marlboro clothes, Camel boots), and sponsorship of high-profile events (such as Formula One) are two industry responses to stronger national regulation of tobacco advertising (see Chapter 5 in this volume). Trademarks and their use in marketing fall within the remit of the TRIPS and GATS agreement (see below), and there is a danger that the agreements could serve to curtail measures to protect public health.

10.4   Agreement on Sanitary and Phytosanitary Measures (SPS)

The Agreement on the Application of Sanitary and Phytosanitary Measures (SPS) deals with issues related to food safety and animal and plant health regulations. The agreement encourages members to base their measures on international standards, guidelines and recommendations where they exist. It also recognizes governments' rights to take sanitary and phytosanitary measures, but stipulates that they must be based on science, should be applied only to the extent necessary to protect human, plant life or health, and should not arbitrarily or unjustifiably discriminate between member states where identical or similar conditions prevail (WTO, 1995).

The relevance of the SPS Agreement relates to Article XX setting general exceptions in the General Agreement on Tariffs and Trade (GATT). The SPS Agreement elaborates issues in relation to measures implemented to protect animal, plant or human life or health (GATT, 1994; WTO, 1994b). Thus, while governments are given rights to institute measures on sanitary and phytosanitary fields, these should respect the stipulations set in the SPS Agreement. This issue has been raised in the dispute settlement of the European ban on hormone-beef and the Canadian appeal against stricter regulatory measures on asbestos in France (Butler and Spurgeon, 1997; WTO, 1997, 1998a,b, 2000a, 2001a).

The European Commission has emphasized a precautionary approach to regulatory efforts and intervened on several occasions emphasizing precautionary principle with a Commission communication on the issue (European Commission, 2000a). The battle, predominantly between Europe and the United States, has continued in the WHO/FAO Codex Alimentarius Commission with the latest debate on the issue being as recent as the 24th session in July 2001 (Codex Alimentarius, 2001a,b). The EU–US debate is further complicated by realization of the costs and consequences of the regulatory requirements of the SPS to developing countries (Finger and Schuler, 1999; Otsuki *et al.*, 2001). The matter of genetically modified products and their labelling is also expected to become an issue within the context of the WTO and the SPS and TBT Agreements (Phillips and Kerr, 2000).

10.5   General Agreement on Trade in Services (GATS)

The basic function of the General Agreement on Trade in Services is to liberalize trade in services (WTO, 1994c). While it permits government regulatory measures, the aim of the agreement is to regulate government action in a framework of liberalization of trade in services. The GATS Agreement does not provide any additional benefits for governments in regulating private sector service providers, but the liberalization of service provision is expected to lead to benefits in broad terms as well as lower costs. The less committment governments have to GATS, the more freedom they retain to regulate their own service sector. In the light of the current knowledge concerning the high costs of privatization of health services and problems in services regulation of private providers (see for example Evans, 1997; Rice, 1997), it is not surprising, therefore, that in practice very few nation states have made commitments in the area of health services. The WTO secretariat has, however, hoped for a future change in this due to increasing implementation of more market oriented reforms in the health sector (WTO, 1998c).

For health policy, there are concerns about the impact of GATS on the freedom by governments to regulate health service provision as a basic public service to meet non-economic goals. Specific problems relate to equity of access to services, as well as the role of non-profit actors. Of particular concern is the need to safeguard public services, domestic regulation and government procurement practices (Government of British Columbia, 2001; Krajewski, 2001; Pollock and Price, 2000). The GATS agreement does have provision to exclude public services, but this is a narrow clause covering 'any service which is supplied neither on a commercial basis, nor in competition with one or more service providers'. In most countries, the provision of health services is from a mixture of public and private providers, nonprofit and for-profit actors, as well as third party payers. This increases the chance that commitments under GATS designed to cover the private sector could affect publicly funded health services provided, for example, on a contractual basis or through grants given to non-profit organizations.

The role of national regulatory measures is of greater importance to trade in services than trade in goods. This has meant that there has been, and continues to be, pressure to renegotiate and strengthen Article VI of GATS on domestic regulation. In GATS, commitments are currently made on a sectoral basis listed under 'national schedules'. However, the inclusion of fast track, cluster or 'horizontal commitments' (applicable to all sectors included in a national schedule), and commitments made bilaterally under the GATS agreement, continue to be debated. Such expansions would extend coverage into other sectors without due consideration of their implications (Sauve and Stern, 2000). If these aims are realized during future negotiations, there are widespread implications for those countries that have not yet made commitments in the health sector. It is also expected that countries will face pressure to include private sector services in, for example, health, education and social services, thus extending GATS coverage over these broad areas.

It is also possible that commitments made in bilateral negotiations or in the context of extending services negotiations to government procurement could have a similar impact, resulting in inclusion of public sector contracts related to health services without due consideration of the implications of this. While prospects of the inclusion of government procurement issues in GATS in the near future have been very limited, the European Commission sees this as an important area due to the large economic share of government procurement, covering about 15 per cent of national budgets (European Commission, 2000b).

The EU has emphasized the role of competition policies in the GATS negotiations. In the context of the GATS this might imply introduction of pro-competitive regulatory measures in other services areas, as has been done in the context of telecommunications negotiations in GATS. In health systems the main issue with respect to competition policies is that requirements for the equal standing of public and private service providers in terms of maintenance, infrastructure costs and accounting have promoted more private sector interests than public sector policies and practices. The British experiences have pointed out the high expenses of the private financing initiative in the National Health Service (Gaffney *et al.*, 1999a,b; Pollock *et al.*, 1999). The negotiation process of the GATS Agreement thus holds the potential to force the emergence of a commercially oriented, service provision environment, resulting in the dismantlement of public health services. These concerns have also been raised in analyses of the implications of the WTO on national policies (Price *et al.*, 1999; Pollock and Price, 2000; Sanger, 2001).

10.6 The WTO and health policy: a case-study of Finland

The underlying principles of the national health system in Finland are based on a commitment to equity of access to services and broad public health policies. The national health system is decentralized with local responsibility for service provision, cross-subsidization from central government on the basis of municipal characteristics and additional financing through a national health insurance system reimbursing, for example, part or all costs of prescription medicines and private health services. While municipalities have responsibility for providing services, there are no regulations on how these should be organized. In the 1980s many municipalities contracted out services to the private sector, but during the recession of the early 1990s, these were de-privatized due to cost constraints. Health policies in Finland have traditionally been based on broad intersectoral measures concerning matters such as alcohol and tobacco, in which public health regulatory measures extend to explicit restrictions of direct and indirect advertising. Alcohol sales are controlled through a state monopoly on distribution and pharmacies have a monopoly on the distribution of pharmaceuticals. The health system thus covers both private and public actors, and public funding and regulatory arrangements with both contractual service provision relationships and third party payer arrangements through the national insurance system.

WTO agreements influence national health policies in a variety of ways. The SPS Agreement is quoted directly by the National Veterinary and Food Research Institute for their work in risk assessment and analysis (Eela, 2001). However, while the need to undertake risk assessment and analysis has changed administrative structures and created new units in the administration, many of the older regulations have remained. The administrative load of updating all health-related regulations to fit WTO stipulations would be substantial for any country and, in practice, has not been implemented except in the context of trade disputes.

In drug policy, the cost of pharmaceuticals has been rising. Whether or not this is directly related to the TRIPS agreement, it is clear that new drugs tend to be more expensive than off patent ones. Reimbursement costs rose during the 1990s at a rate of 8–16 per cent per annum with few exceptions (Klaukka and Rajaniemi, 2001). For example, National Sickness Insurance paid 4030 million marks (approximately 680 million ECU or US$610 million) of pharmaceutical reimbursements in 2000, almost 11 per cent more than the previous year. The proportion of the population receiving reimbursements remained constant, so increased costs were mainly due to changes in treatment. In Finland a policy to cut costs during the early 1990s led to larger patient contributions to pharmaceutical costs compared to other European countries. Thus, increases in public spending on drugs are expected to be even higher in countries with more generous reimbursement policies.

In health services, previous negotiations on GATS did not result in Finnish commitments on health services. Finnish state monopolies in the distribution and sale of pharmaceuticals and alcohol were retained, as well as specific commitments on social insurance due to arrangements with employer contributions to pension funds. Thus, national policy has tended to keep health and social services outside GATS commitments. However, as part of the current European preparatory process to GATS negotiations, all service areas are to be reviewed for potential inclusion. Ministries are required to feed their views into European negotiations and, if necessary, make a case to keep existing restrictions. As health services in Finland are provided on a contractual basis by local authorities, it is unlikely that they would automatically be excluded under the WTO public services exception (see above). The situation could become even more open if either OECD-initiated regulatory reforms or EU stipulations on government procurement obligate public services to undertake contract bidding. Another area of possible concern is the practice of subsidizing NGOs through public resources. This is a matter already debated in the context of competition policy and national lottery fund support for Finnish NGOs active in social policies and service provision (Ruohonen, 1999).

In addition to areas directly related to the health sector, other policy areas relevant to health need careful consideration. In GATS insurances are dealt with under financial services. Policies related to public transport and utilities, for example, are the subject of several European Union proposals (WTO 2000b,c, 2001b,c). In Finland, water utilities are mainly public enterprises, with supplies metered and charged. The sector could readily be covered under WTO commitments without due consideration of its consequences, as water utilities and services are dealt with under environmental services. Another problematic area is advertising, where Finland has already made broad commitments. However, two recent decisions by the European Court of Justice highlight the problems of these commitments – rights of access of advertising services counteracted policies restricting alcohol in Sweden and tobacco advertisement at the European level (European Court of Justice, 2000, 2001). The same argument could be raised to counter restrictions on any advertising on the grounds of public health, such as Finnish law regulating indirect advertising of tobacco products.

Ministries of health are generally weaker in comparison to Ministries of Trade and Industry or Ministries of Foreign Affairs, and as a consequence have limited power in setting national priorities. This 'health policy deficit' becomes worse at the European level. In the European Community, subsidiarity principle has kept the sphere of health and social services in the national remit. The European Community level policies have focused on promoting internal markets, trade, competition and industrial policies, thus easily resulting in lack of recognition of the influence of these on health and social policies. The recent developments in the European Community have also tried to shift more power and capacities to European level decisionmaking, with the latest decisions taken in the intergovernmental conference in Nice (European Commission, 2001).

As health and social services issues are not in the sphere of EU competence, matters related to these sectors are not actively considered at the European level, resulting in bias towards services industry and export viewpoints in these areas. In addition, some policies, such as pharmaceutical policies, are mostly dealt with in the context of industrial policies at European level, further complicating the articulation of health policy viewpoints at European level. It is also known that the pharmaceutical industry is one of the major lobbying forces at European policy level (Greenwood, 1997).

In the European Community, trade policy is mainly dealt with by Committee 133, primarily consisting of foreign and trade policy experts. In Finland inputs are provided by a national background group, which carries out consultations with relevant ministries and civil society groups on such policy areas as the environment, agriculture, labour, health and education. In practice, however, this consultative process is problematic due to the short time-period available to review Committee 133 documents. This means that the European Commission does a large share of the negotiations and background work for member states, even when national parliaments, in principle, have the last word in accepting policies.

Overall, the brief discussion above suggests the need to go beyond blanket assumptions that enhanced trade through WTO agreements will lead to economic growth and, in turn, positive health effects (Feachem, 2001). This is the argument that continues to drive the trade agenda, with health tacked on as an optional extra. Each trade agreement needs to be understood in terms of its direct and indirect impacts on health, and other sectors related to health. More fundamentally, health and other social policies cannot be reduced to arguments centred on economic growth, but must include considerations of distributional matters, social equity, quality of life and work and social cohesion.

10.7   Future challenges for strengthening health policy amidst globalization and trade

This chapter discusses various ways in which WTO agreements can influence the nature of health policies. Debates on health and MTAs have so far been largely polarized between pro- and anti-globalization perspectives, with limited analysis of the specific distribution of costs and benefits from trade among the wide range of public and private actors in the health sector. In recent times, the impact of MTAs on access by low-income countries to essential drugs has received considerable attention, and similar study of other substantive areas is much needed. This section considers three aspects of policymaking that require fuller assessment in order to improve understanding of the implications of MTAs for health: (a) broader processes of policy change and anticipating future change; (b) emerging forms of global health governance; and (c) priority setting at the local, national and global levels.

Any assessment of future negotiation priorities must also predict future changes. Understanding changes taking place over the next ten to 20 years is necessary in order to make an appropriate assessment of matters of concern related to health services. For example, it is clear that the Internet and telemedicine have enlarged the prospects of transnational trade in services in ways that were not easy to foresee ten years ago. In the debates on implications of trade agreements it is often only the critical viewpoints which are required to present evidence, although during the negotiation process all implications, and expected benefits, are based on expected and possible implications rather than actual facts and experiences.

WTO Agreements may, in many policy areas, have significant impacts even though the actual policy changes that occur may be driven by other processes and only later legitimized with reference to the WTO. In some cases, the actual policy change is driven by other institutions, such as the World Bank, IMF and OECD, and later legitimized through the WTO (Anon, 1996; OECD, 1997). The OECD has been an especially important promoter of health care reform in high-income countries, and has acted as a mediating forum for the globalization of such reforms (Moran and Wood, 1996). The WTO Agreements can also be used by national corporate actors to carve out public benefits or limit regulatory efforts. It is thus necessary to see the domestic elements in the reforms conducted as technical and efficiency oriented rather than political and international processes. Health sector reforms introduced at the national level, in such forms as deregulation, privatization and decentralization, are restructuring health systems to make them complementary to the emerging global economy.

A recent study of future health care markets in India, for example, locates anticipated change within broader macroeconomic reforms and the influence of international financial institutions (Shekri, 2001). This role of international financial institutions is also clear in the work of the International Finance Corporation (IFC), the private sector financing arm of the World Bank group, which maps prospects for private investment in health care in various countries and actively supports private sector activities (Taylor Associates International, 1997). The World Bank and IMF corporate-friendly influence in Latin American has also gained critical analyses of their influence on Latin American health systems (Stocker *et al.*, 1999). WTO agreements can thus serve to lock in reform processes initiated bilaterally and nationally, giving the impression of little change from WTO agreements. However, it is not necessarily the magnitude of change that matters, but the difficulties of reversing policies after they are agreed multilaterally. It is easier to change national or bilateral policies than multilateral agreements.

As well as dealing with processes of policy change, there are challenges arising from the governance of health globally. The increased interconnectedness across sectors of policy issues raises questions about the appropriate mandates of different international organizations. In the context of this chapter, there is a clear and growing need to deal with conflicts between trade and health policy goals. From the perspective of health policy, it would seem appropriate for WHO to deal with issues of public health relevance rather than the WTO. The latter's authority to enforce compliance with its agreements has led to arguments in favour of relocating substantive issue areas under its remit, including the biodiversity convention, pharmaceutical regulation and labour rights. This has been the aim of many NGOs and trade unions, although opposed by many countries as beyond the trade sphere. However, UN specialized agencies such as WHO, UNESCO, FAO and ILO currently have the mandate to deal with many matters being proposed for the WTO. Rather than expanding the WTO's mandate, it may be preferable to make WTO policies more accountable to existing forms of governance, including UN organizations, international agreements and recognized principles protecting human rights and the environment.

The danger is that in many cases other organizations, such as UN specialized agencies, have already been incapacitated by shifts in the negotiating frameworks on important substance issues. Braithwaite and Drahos (2000) see this forum-shifting process encompassing three kinds of strategies: moving the agenda from one organization to another, abandoning an organization and pursuing the same agenda in more than one organization. They also define forum-blocking, where a powerful state ensures that an international organization does not become a forum for an agenda that threatens its interests. They present an example of UNESCO and copyrights, and cite the abolition of the UN Centre on Transnational Corporations as the most extreme example of forum shifting. In the sphere of health policies parallel examples are apparent in the role of the WHO in pharmaceutical and regulatory policies, where the EC initiative establishing the International Conference on Harmonization became an important forum in the globalization of pharmaceutical regulation because of the support of the US, the EC and Japan (Braithwaite and Drahos, 2000). In the case of health policy issues, NGO campaigns that emphasize WTO capacities for enforcement may increase the chance of health matters shifting to the G8, the World Bank and the WTO as part of the same process.

In trade policy there are clear potential conflicts between public and private interests regarding regulation. While the public interest behind standards-setting focuses on the protection of health and safety, for example, private interests seek standardization as a means of facilitating access to markets. Different international organizations responsible for standards reflect such diverse interests, ranging from WHO to corporate funded organizations such as the International Standards Organization (ISO). WTO stipulations do not specify the appropriate type of funding mechanism for standard-setting organizations. This raises the concern that standards-setting to support public policies is undermined by institutions serving corporate needs. Ironically, while trade liberalization is often presented as benefiting consumers, the formal opportunities for representing consumer interests in standards-setting are increasingly difficult. In short, the nature and membership of standards-setting organizations needs to be carefully considered to ensure that there is an appropriate and accountable balance between public and private interests.

The third challenge concerns how priorities are set among different sectors and at various levels of the health policy process. As noted above, health policy at the national level generally holds relatively low status compared to policies of finance, trade and industry, and foreign affairs. As such, national positions during trade negotiations tend to derive from the perspectives of the latter, accompanied by strong representation from corporate interests seeking import–export markets. It is difficult, in turn, for the concerns of social welfare ministries to contribute to the trade agenda. Nonetheless, trade liberalization and its accompanying policies cannot be seen as ends in themselves, but means towards broader societal aims such as improved quality of life and human welfare.

The capacity of traditional public health organizations to engage effectively in trade negotiations at the regional or global level is also a challenge. While WHO work on access to drugs has been pathbreaking (Velasquez and Boulet, 1997; WHO, 1998h, 1999c), concerted and broad ranging attention to trade issues remains to be developed. In part, this is due to a lack of specialist expertise on the health implications of international trade law, hindering the ability to act even when there might be scope to do so.

Issues of power and representation have been raised in the context of developing countries and WTO negotiations, but these should also be considered in relation to health policies. The deficit of health policy viewpoints can be clearly seen in the European processes described earlier as part of Finnish health policy issues at national and European level. Prior and Sykes have compiled and analysed the implications of globalization on welfare states and one of their conclusions is an emphasis on regionalization and regional implications, especially among European Union member states (Prior and Sykes, 2001). In the case of the Finnish policies and the WTO, it can clearly be seen that many of the implications relate to the regional context rather than directly to global level.

Many of the health issues arising under WTO agreements are highly specialized, requiring detailed understanding of technical and legal terminology. This is mirrored at the national level where NGOs and other civil society groups may be vocal on health issues, but suffer from similarly weak capacity. Relatedly, individual organizations can also be too narrowly focused on specific drugs, diseases or treaties, yet unable to speak authoratively on the health implications of the emerging trading system as a whole.

10.8   Concluding viewpoints

Trade negotiations tend to be conducted from the perspective of trade benefits and with strong representation of corporate and export interests.

This bias leads to problems in other areas of public policies with competing and contrasting aims and which remain unrecognized and inadequately addressed. It is thus important that implications of WTO Agreements are assessed in terms other than trade perspectives and that appropriate analysis and review is required before further negotiations. Economic integration or protection of intellectual property rights are not aims in themselves, but should be means towards supporting broader societal aims.

As a global rule-setting agency the WTO is too often considered in terms of competing national interests in contrast to more systematic analysis on how the WTO agreements relate to commercial and corporate rights on one hand and citizens rights, principles of solidarity and public policy on the other. Health is only one of the substance matters of concern; other areas are environment, culture and education. However, the strong national basis of health policies easily obscures international aspects. Health policy actors and policy forums are not either sufficiently prepared or empowered to raise health matters higher on the political agenda or to seek broader consideration of health policy priorities in the trade-related political processes remaining largely invisible to decisionmakers. This results in a deficit in health policy at local, national and global level.

Finally, international trade agreements are not based on natural laws nor represent holy texts that must be kept as they were written. The alternative is not global trade policies without any rules or protectionism, but rather the adjustment of rules to respect and provide for policies that promote human and social rights and socially and environmentally sustainable societies.

11

Trade Policy, the Politics of Access to Drugs and Global Governance for Health

Caroline Thomas

> Never have so many had such broad and advanced access to healthcare. But never have so many been denied access to health.
>
> (Gro Harlem Brundtland,
> the Director-General of WHO, December 1998)

> Access...amounts to a moral problem, a political problem and a problem of credibility for the global market system.
>
> (Brundtland, 2000)

11.1 Introduction

Significant advances have been made in global health over the last 50 years; for example, life expectancy has increased from 48 years in 1955, to 66 years in 1998. However, we cannot overlook the fact that these advances are 'marred by growing health disparities between the world's wealthy and the world's poor' (Millen *et al.*, 1999, p. 4). Nowhere are these disparities seen more clearly than in the experience of access to drugs.

In Spring 2001, the issue of access to drugs was catapulted onto the global political agenda as a transnational alliance of NGOs stepped up their campaign to widen access to anti-retroviral (ARV) drugs for HIV/AIDS sufferers worldwide (see below). These NGOs have argued that the efforts of a few developing countries to pursue legitimate strategies to secure drugs for their people at affordable prices have been obstructed by the combined might of the pharmaceutical industry and the US government.

This chapter explores the relationship between trade policy and access to drugs, using the anti-retroviral (ARV) drugs as an example. It begins by noting the moral problem of inequality in access to drugs. It goes on to explore the political problem in terms of the discrepancy between what is

Box 11.1 The moral problem: inequality in access to drugs

'The inequalities are striking. In developed countries, there may be one pharmacist for every 2000 to 3000 people. A course of antibiotics to cure pneumonia can be bought for the equivalent of two or three hours wages. One year's treatment for HIV infection costs the equivalent of four to six months' salary. And the majority of drug costs are reimbursed.

In developing countries, there may be only one pharmacist for one million people. A full course of antibiotics to cure common pneumonia may cost one month's wages. In many countries, one year's HIV treatment – if it were purchased – would consume the equivalent of 30 years' income. And the majority of households must buy their medicines from their own pockets.'

Source: J. A. Scholtz, Executive Director, Health Technology and Pharmaceuticals, WHO, 'Views and Perspectives on Compulsory Licensing of Essential Medicines', 26 March 1999, http://www.haiweb.org/campaign/cl/scholtz.html.

legal/permissible under WTO rules and what is permissible/desirable under the terms of US trade policy. Finally, it explores the problem of credibility for global health governance and the global market system, in terms of access to drugs.

Perhaps more than any other disease, HIV/AIDS reflects entrenched and growing global inequality and exclusion, and the continuation of the North–South divide. The problem is overwhelmingly (95 per cent) a problem of the South; in particular, it reflects the continuing marginalization of Africa (Booker, 1999). About 35 million people worldwide are HIV positive; 26 million of these are African. HIV/AIDS remains incurable, although with appropriate combinations of drugs quality of life can be improved and life expectancy considerably enhanced.

The results of the disease are very different, depending on whether one's fate is to be born in the developed or developing countries.

> Today, hundreds of thousands of people with the disease in the industrialized world lead full, healthy lives, thanks to (ARV) drugs. In the developing world, perhaps only one in a hundred of those needing treatment have full access to ARVs. The vast majority of people living with AIDS in the developing world receive either no medical treatment or only palliative care to reduce pain and suffering. (Panos, 2000, p. 3)

Of those HIV infected, Panos estimates that 12 million in the developing world need ARV drugs now. The overwhelming majority will not get them. Moreover, even if the drugs were available, an appropriate infrastructure would need to be developed for their delivery. In a global environment where aid commitments have fallen over the last decade, and where debt reduction has been slow in coming, the costs of such infrastructural development are prohibitive.

In 2000, the Panos Institute, London, undertook a thorough study of the costs of treating HIV/AIDS (Panos, 2000). It estimates that worldwide, US$60 billion a year was needed at 2000 prices to pay for ARVs, and that this would rise (Panos, 2000, p. 1). To put this figure in context, this is equal to less than 25 per cent of the US annual military budget. It is US$8 billion more than the amount annually spent on obesity in the US (Piot, quoted in UNAIDS/WHO press release, 28 November 2000). In the case of a country like Zambia, to buy the necessary drugs at current prices for those who need them would cost US$2 billion, equivalent to 57 per cent of Zambian GDP. Panos has estimated that in the developing world, it costs roughly US$4000–6000 per person per year for a course of ARV drugs, and the associated tests and consultations (Panos, 2000, p. 3).

Despite the denials of pharmaceutical companies, the fact is that *differential access to ARV drugs due to cost contributes to the uneven global experience of HIV/AIDS.* These drugs are produced largely, but not wholly, in the North and many of them are under patent. Some of these drugs are not on the WHO list of Essential Drugs, because they are too expensive. A few developing countries, such as India and Brazil, have the ability to produce generic versions of some of these patented drugs. These are much cheaper than their patented cousins. The price of patented drugs puts them out of reach of the overwhelming majority of sufferers. The African Development Forum in Addis Ababa, December 2000, mindful of this, argued in its final declaration: 'A substantial reduction in the prices of antiretroviral drugs and treatments for opportunistic infections is required. African governments, donors and international financial institutions must work in partnership to reduce the price of drugs to a level commensurate with production costs' (African Development Forum, 2000).

For a concrete example of price differentials between patented drugs and generic cousins, let us consider the example of fluconazole, which is used to treat cryptococcal meningitis, among other things. Ten per cent of people with AIDS suffer from this; in some areas, 25 per cent. Without treatment, life expectancy is one month. The drug is under patent to Pfizer until 2004 in the US. However, since not all countries recognize patents on medicines, it is also being produced generically elsewhere. MSF reported in *The Lancet* on 16 December 2000 that the company has refused to grant voluntary licenses to enable poor countries to import an affordable generic supply. If South Africa imported generic fluconazole to treat this problem, this 'would have a striking effect on access and adherence to treatment' (Perez-Casas *et al.*, 2000, for MSF). If the South African government imported generic flucanazole from Thailand, the price of yearly treatment would be reduced from the current US$2970 to US$104.

Table 11.1   Comparative study of generic and patented flucanozole: wholesale prices of 200 mg capsules, June 2000

Manufacturer	Country of production	Country of distribution	Price per unit (US$)
Biolab	Thailand	Thailand	0.29
Cipla	India	India	0.64
Bussie	Colombia	Guatemala (negotiated)	3.00
Pfizer		Thailand	6.20
Vita	Spain	Spain	6.29
Pfizer		South Africa	8.25
Pfizer		Kenya	10.50
Pfizer		Spain	10.57
Pfizer		Guatemala (negotiated)	11.84
Pfizer		US	12.20
Pfizer		Guatemala (not negotiated)	27.60

Source: Carmen Perez-Casas *et al. HIV/AIDS Medicines Pricing Report*, Médicins sans Frontières, July 2000, accessed at: www.mst.org/advocacy/accessmed/reports/2000/07/aidspricing/index.htm.

11.2   The political problem: the politics of access to ARV and other drugs

Since the late 1990s, a small number of developing countries, with the support of a transnational alliance of NGOs, have been battling for affordable access to essential ARV drugs. It has been estimated that the cost of ARVs will have to be reduced by 95 per cent before they can be affordable to all that need them (Panos, 2000, p. 2). These efforts of a few developing countries to pursue legitimate strategies to secure drugs for their people at affordable prices have been obstructed by the combined might of the pharmaceutical industry and the US government. Interestingly, these developing countries have been fighting only for what they are legally entitled to under the WTO Trade-Related Intellectual Property (TRIPS) agreement: the use of compulsory licensing and parallel importing to increase access to affordable drugs for their infected citizens. Richard Laing (1999) argues that manufacturers of proprietary drugs would not be affected in any significant way by changes in pricing such as compulsory licensing, as the proportional contribution of Asia, Africa and the CIS is so small to both turnover and profit, of these pharmaceutical giants. However, the manufacturers of the ARV drugs do not agree.

11.3   WTO/TRIPS and access to drugs

The issue of patent protection has been high on the international trade agenda since the establishment of the WTO in 1995. The TRIPS agreement sets a minimum standard for intellectual property protection in all member countries' national legislation. In the case of pharmaceuticals, patent protection is extended for a minimum of 20 years. Developing countries had until 2000 (or 2006 for the least developed) to bring their national policies into line with this.

In theory, the TRIPS does allow countries to protect public health; however, in practice this does not always seem to have been the case (see below). Under Article 8.1, TRIPS says that: 'members may ... adopt measures necessary to protect public health and nutrition, and to promote the public interest in sectors of vital importance to their socio-economic and technological development'. Under certain circumstances, TRIPS allows countries to pursue parallel importing (Article 6, Exhaustion of Rights) and compulsory licensing (Article 31).

Parallel imports refer to importing a patented drug from a third party in another country where it is sold for less. Under Article 28 of the TRIPS, patent owners have the right to prevent third parties from 'making, using, offering for sale, selling or importing' a product, but it is states who determine when these rights are 'exhausted'. Under Article 6, states can take whatever action they deem necessary at the point of exhaustion. This allows for parallel imports as national policy, which is permitted under EU, US and Japanese patent laws (Love, 1999a).

Compulsory licensing permits the manufacture (anywhere) and use of generic drugs without the agreement of the patent holder. Under Article 31 of WTO/TRIPS rules, states can issue such licenses for a number of reasons, not only national emergencies, so long as they adopt adequate safeguards such as compensation. In such emergencies, however, as in the case of non-commercial public use, or to correct anti-competitive practices, they do not need to make prior efforts to negotiate a license on reasonable commercial terms with the patent holder (Love, 1999a).

Under WTO rules, decisions regarding the appropriate amount of compensation paid to patent holders are decided under the national law of the country issuing the license. Those national laws determine the ability of the country to import drugs via compulsory licenses. Love points out that

> the TRIPS does have some limits on the ability of a country to export under a compulsory license, but drugs can be acquired from non-WTO member countries, and from WTO member countries where the drug is off patent, or where exports are not the predominant activity, or in countries that provide patent exceptions for imports into countries that have TRIPS compliant compulsory licenses. (Love, 1999a)

Only a small number of developing countries have the medical and industrial infrastructure to produce these drugs themselves and also a stratum of the population able to purchase them. As Wright points out, these are the very countries in which pharmaceutical companies would like to expand their market (Wright, 1999, p. 4). These countries have found that going down this road elicits a very heavy-handed response from the US. Indeed, one author has commented that: 'compulsory licensing ... is a dangerous weapon, in terms of generating a very dramatic response' (Laing, 1999, p. 3).

11.4 US power and the issue of access

The US argues that the TRIPS is the minimum standard acceptable for patent rights and in its bilateral dealings it encourages other countries to go for more than the minimum required under international law. This position was clarified by Lois Boland of the US Patent and Trademarks Office in the Geneva conference on compulsory licensing in March 1999: 'In our bilateral discussions, we continue to regard the TRIPS agreement as an agreement that establishes minimum standards for protection and, in certain situations, we may, and often do, ask for commitments that go beyond those found in the TRIPS agreement' (Boland, 1999).

The US government has successfully put Thailand under pressure to change its patent and trade laws so that they are more restrictive than what is allowed under TRIPS. James Love, of the Consumer Project on Technology (CPT), Washington DC, comments:

> The problem for developing countries is not whether compulsory licensing of pharmaceuticals is legal, because it clearly is legal. It's the political problem of whether they will face sanctions from the United States government, for doing things that they have a legal right to do, but which the United States government does not like. In the case of Thailand, that country clearly could have done compulsory licensing on these drugs for meningitis and AIDS. They had a statute in place that gave them the authority to do it, and it was consistent with international law. But the US government threatened trade sanctions, and used a carrot and stick approach to persuade the Thai government not to do something which would have been legal under international law. (Cited by James, 1999b)

The US has been less successful in the case of South Africa. South Africa became the focus of a bipartisan US campaign to get it to amend or repeal the Medicines and Related Substances Control Amendment Act ('Medicines Act') of 1997. Ralph Nader and James Love of CPT have referred to the 'weight of US power, short of military warfare, on South Africa to prevent that country from implementing policies to obtain cheaper sources of essential medicine' (cited by James, 1999b, p. 4).

The Medicines Act was passed by the South African Parliament in 1997. However, it was challenged in the local High Court by over 40 pharmaceutical companies, who claimed it was unconstitutional. The dispute was stuck there until April 2001 (see below). A key aim of the act was to enable the government to purchase generic drugs at affordable prices. The health system was undergoing major reform, with the right to health care for all being constitutionally embedded in 1996. This made the issue of drug prices all the more important (Bond, 1999, p. 767).

The Medicines Act included a raft of provisions for increasing access to affordable drugs. However, the one which most offended the US was Clause 15(c):

> ... The Minister may prescribe conditions for the supply of more afford-able medicines in certain circumstances so as to protect the health of the public, and in particular may...prescribe the conditions on which any medicine which is identical in composition, meets the same equal-ity of standard and is intended to have the same proprietary name as that of another medicine already registered in the Republic...may be imported. (Cited in Bond, 1999, p. 768)

The US objected to the legitimization of parallel importing and compulsory licensing. Leon Brittan of the EU wrote to South African Vice-President Thabo Mbeki in support of the US position, claiming that South African laws were at variance with WTO obligations and that EU companies would be hurt by this action (Taylor, 14 March 2000).

South Africa was punished for not coming into line by being put on the US trade Special 301 Watch List in April 1998. In June 1998 it was denied Generalized System of Preferences (GSP) treatment for four items, pending progress on intellectual property protection. In 1999, the pressure intensi-fied when US Trade Representative (USTR) Charlene Barshefsky, citing South Africa's advocacy role in the World Health Assembly, called for an 'out of cycle' review of South Africa to be held in September 1999 (Bond, 1999, p. 776). The 30 April USTR 301 report on South Africa claimed that South African representatives 'have led a faction of nations in the WHO in calling for a reduction in the level of protection provided for pharmaceuti-cals in TRIPS' (www.ustr.gov/releases/1999/04/99-4.1html). As leader of the Non-Aligned grouping at that time, South Africa was well placed to give the issue of access to medicines greater importance on the World Health agenda, and to increase support globally for this. At no time, however, did it call for a change in the TRIPS. At the 52nd World Health Assembly, January 1999, a unanimous resolution was passed which gives health a place in trade negotiations. The US eventually backed down *vis à vis* South Africa and ended trade pressures (Love, 1999a). In May 2000, South Africa was removed from the trade 'watch list'.

Why does the US take this stand against various methods for making drugs more affordable? One reason, of course, is that pharmaceutical lob-bies in the US are incredibly powerful. Consider that promotional spending by companies in the US in 1997 was US$4.2 billion, equivalent to the total drug sales in Africa (Laing, 1999, p. 3). And that Pfizer has more staff in its marketing department than the whole of the WHO (Koivusalo, 1999, p. 38). These companies exert pressure on the US government to promote and defend their interests abroad. This is not new.

The case of Bangladesh in the early 1980s is an infamous example of US foreign policy serving the interests of pharmaceutical companies, rather than public interest in broad access to health. The efforts of the Bangladeshi government to streamline spending on drugs by use of a list of essential drugs met with a very hostile reaction from the companies, who urged the US government to encourage the government of Bangladesh to change its mind. This it did.

The US government even includes representatives of the industry in its official visits to other countries. In 2000, for example, the president of Merck joined US State Department officials on a visit to Brazil, the purpose of which was to encourage the Brazilian government to abandon legislation that would increase access to affordable AIDS medications.

US policy seemed to be changing in May 2000, following NGO campaigning on the issue in the election year. President Clinton signed an Executive Order which 'prohibits the US Government (from bringing trade sanctions) with respect to any law or policy in beneficiary sub-Saharan countries that promotes access to HIV/AIDS pharmaceuticals or medical technologies and provides effective and adequate intellectual property protection consistent with the TRIPS agreement' (James, 2000). In other words, the US would accept the WTO standard on patents, rather than requiring more stringent US trade law standards. While this represented a significant change, critics asked why the step was limited to Africa and to AIDS drugs (see Kaiser Daily HIV/AIDS Report, 2000a). The AIDS problem extends to other continents, and within Africa there are many other important health challenges such as TB and malaria.

The industry response was hostile. Alan Holmer, the president of the Pharmaceutical Research and Manufacturers of America, commented that the Executive Order sets 'an undesirable and inappropriate precedent, by adopting a discriminatory approach to intellectual property, and focusing exclusively on pharmaceuticals' (Kaiser Daily HIV/AIDS Report, 2000a).

The situation has been evolving rapidly. President Bush in January 2001 considered reversing Clinton's Executive Order, so the WTO/TRIPS standard may not satisfy US trade representatives in future (Kaiser Daily HIV/AIDS Report, 2001). In terms of equal access to drugs, this would be bad news. Also worrying is that the US raised the issue of Brazilian patent laws at the WTO, and asked for an arbitration panel to investigate their conformity with WTO rules. Despite all the US protestations against Thailand and South Africa, it did not take a dispute to the WTO for adjudication. One possible reason was the calculation that it probably would not win. However, with a new government and new relationships with pharmaceutical companies, policy has been shifting and muscles have been flexed. Given the Brazilian success in producing ARV drugs since 1996 and treating patients free of charge, this is potentially a huge blow not only to AIDS sufferers in Brazil, but worldwide. Brazil has offered these drugs to other countries. NGOs such as Oxfam and Médecins Sans Frontières have immediately gone into top gear, campaigning against this. The implications of the case go far beyond HIV/AIDS (http://www.oxfam.org/cutthecost/) (also, see ACT UP Paris press release, 2 February 2001, 'The WTO Menaces the Survival of 100,000 People with AIDS'). The findings will set a precedent.

11.5 The role of the pharmaceutical giants in issues of access to drugs

The powerful pharmaceutical companies in the North vehemently oppose attempts by developing countries to produce or acquire cheap drugs, especially via methods that would be most likely to result in a sustainable solution. They are opposed to generic production in the South, even though many of them are involved in it themselves in the North (Nogues, 1990). As we have seen above, they are supported in this stance by the US government, which has pursued a very aggressive trade policy to ensure a very strict definition of international patent protection (Wright, 1999).

Companies can and do pressure states and generic drug producers directly, to persuade them to change their policies, even if they are acting in accordance with international law. One recent example involves Glaxo Smithkline (GSK) and Ghana. GSK has put Ghana, the 5 per cent of its population HIV infected, and the Indian company Cipla under enormous direct pressure. Indirectly this pressure, in the form of a clear signal, extends to other exporters and potential importers and users of generics.

Cipla has been exporting low cost generic Duovir (AZT and 3TC) to a Ghanaian drugs distributor, Healthcare Ltd. Glaxo has accused Cipla of patent infringement, by violating Glaxo's Combivir (brand name for 3TC and AZT combined in one pill) patent rights. In August 2000, GSK threatened to take Cipla to court.

However, it seems that Glaxo's patent rights are not valid in Ghana (Schoofs, 2000b). The patent system is not retroactive and Ghana did not allow patent protection for pharmaceuticals until 1 July 1993. The Global Treatment Access campaign argues that: 'because GSK filed for several patents relevant to Combivir before Ghanaian patent law recognized patents on medication, it is likely that GSK's claim to patent rights to Combivir are completely invalid' (www.globaltreatmentaccess.org/content/camp/...ghana.htm dated 30 November 2000). In other words, GSK has no exclusive rights to market the medication in Ghana. At least three out of four of Glaxo's patents on Combivir are prior to July 1993. Cipla's and Ghana's actions were lawful.

Yet Glaxo's tactics have been successful; Cipla ceased exports to Ghana and Healthcare Ltd is afraid to distribute the drugs that have already reached Ghana. In the meantime Ghana's HIV sufferers continue to die, while Glaxo continues to negotiate price reductions for Combivir through the UNAIDS initiative. Here we have the largest pharmaceutical company in the world (7.3 per cent of the global market, with control of over a third of the ARV market – www.globaltreatmentacccess.org/120100_HG-GSK_GHANA.htm 1 December 2000 'GSK Profiting from Barriers to Essential Medicine'), bullying a poor country as well as a small company, neither of whom were breaking international law.

Why are companies so opposed to generic production, parallel importing, compulsory licensing and so on? The most frequently cited reason is that financial incentives are necessary for research and development (R&D). Companies claim that R&D costs involved in developing new drugs are so high that developing countries are ill advised to turn to generics, as this will discourage the transnationals from further R&D to deal with Southern diseases. On the surface, this argument is very compelling. If companies do not have lengthy patent protection, they do not have an incentive to pour resources into the development of new drugs. For a number of reasons, this argument is highly questionable.

First, if it were really the case, many of the developments of the last five decades would not have occurred through lack of patent incentive. For it is only in recent decades that many developed countries themselves developed patent laws on drugs (the older US patent laws are an exception). Their pharmaceutical industries for the most part developed without this kind of protection (Challu, 1991, pp. 74–7; Nogues, 1990, pp. 82–3). Challu argues that: 'most industrialized countries adopted product patent protection systems once they had already reached a high degree of economic development' (1991, pp. 74–5). And further that 'the hypothesis that increased patent coverage encourages more invention may be regarded as false, based on empirical evidence from the US, as well as on a world-wide level' (1991, p. 86).

Second, R&D priorities are set by companies not according to public health needs, but rather according to calculations about maximizing the return to shareholders. Developing countries do not represent a lucrative market. The global pharmaceutical market is huge – over US$400 billion per annum. Yet Africa accounts for only 1.3 per cent of the global health market. About 90 per cent of the US$70 billion invested annually in health R&D by pharmaceutical companies and Western governments is not focused on tropical problems, but increasingly on the problems faced by the 10 per cent of the global population living in developed, industrialized countries. Examples include baldness and obesity. It explains why of the 1233 new drugs that entered the market between 1975 and 1997, only 13 were targeted specifically at tropical, infectious diseases (Pecoul, cited by HAI, May 2000 at http://www.haiweb.org/news/WHA53en.html).

In their advertising, companies can be somewhat misleading on this point, as demonstrated by the advert below announcing the union of two large companies.

> Today is the day. Today is the day 139,000 people will die prematurely from disease. Over 25,000 of them will be children under five. But today is also the day that Glaxo Wellcome and SmithKline Beecham become one. This means that for the first time, over 100,000 people will pool their unique talents to seek causes and find remedies for diseases all over the world. They will do so not just with a sense of hope. But with a sense of urgency. Diseases do not wait. Neither will we. (GlaxoSmithKline advert in *The Economist*, January 2001).

But they have not poured resources into finding cures for diseases of the poor. Indeed the Global Forum for Health Research reported in 2000 that less than 10 per cent of global health research funding was allocated to 90 per cent of the world's health problems, mostly concentrated in developing countries (World Bank, 2000).

Third, much R&D is initially paid for by Northern taxpayers. The US government funded much research into tropical diseases such as malaria when it had troops in active service abroad. In the case of ARV and related drugs, the US government, through the National Institutes of Health, has funded primary research. While it is true that pharmaceutical companies spend a significant amount on R&D, one should not forget the high profits they make.

An alternative explanation of the companies' dislike of generic production in the South, is that they are acting to ensure that they can continue to protect market share in developed countries and continue charging high prices there. This is especially so in the face of increased competition from generics and tighter drug safety and efficacy regulations in the North (Nogues, 1990, p. 81). Ultimately they are protecting their profits.

Companies claim that the price of a drug reflects among other things, the costs of R&D. Critics, however, claim the prices of proprietary drugs often reflect what the market will bear rather than the costs of R&D. Duckett illustrates this with the case of Pentamidine, a cheap treatment developed for sleeping sickness. She points out that when this drug was found to be effective in treating AIDS-related PCP (*pneumocystis cariniii pneumonia*), the price increased 500 per cent and it evaporated from the market in poor African and Southeast Asian countries (Duckett, 1999, p. 5).

11.6   The credibility problem: the market system, public–private partnerships and the challenge of widening access to drugs

During 2000, the success of activists in politicizing the issue of access, coupled with bad publicity for pharmaceutical companies and also the electoral imperative in the US, prompted some new initiatives. Importantly, however, patents continued to be supported over cheaper forms of drugs.

In July 2000, President Clinton announced US$1 billion Export–Import Bank loans for the import of drugs at patented prices. This met with sharp criticism from campaigning groups and Southern countries, which saw this as adding to the already unbearable burden of debt of the poorest countries.

Potentially more significant was the joint UNAIDS–five company initiative announced in May 2000, and it is to this public–private partnership that we now turn. The partnership between five pharmaceutical companies and UNAIDS aims to bring ARV and related drugs to people in the South who cannot afford to buy them in the market place. We turn to this partnership to help us establish the efficacy of such arrangements to deal with the challenge of access to drugs.

The initiative built on the far more limited version of 1997 that aimed at bringing reduced priced drugs in limited quantities to four countries (Vietnam, Senegal, Uganda and Côte d'Ivoire). On paper the 2000 version represented a significant scaling up of the attempt to bring more drugs more cheaply to more countries and people. Amidst great fanfare, the companies announced their intentions to the *Washington Post*. However, they omitted any details of how much, for how long and for whom.

Details published in the *Washington Post* about how the initiative was agreed are quite shocking, revealing a startling lack of partnership between the UN agencies and the companies in negotiating the deal, and a lack of clear vision regarding policy and sustainability (Gellman, 2000b). There was no involvement of Southern governments or concerned civil society groups.

Box 11.2   Six month report card: UNAIDS–five companies accelerated access programme

- Number of countries that have negotiated price reductions to date (December 2000) – one (Senegal)
- Number of people with HIV in sub-Saharan Africa – 26 million
- Number of patients that will benefit in Senegal once this programme is implemented (according to UNAIDS) – approximately 900 (out of 79,000 with HIV)
- Number of patients that Brazil has put on anti-retroviral therapy by using affordable generic medicines – more than 90,000
- Amount of money Brazil has saved on hospitalizations and treatments for opportunistic infections avoided by successful use of anti-retroviral therapy (1997–99) – US$472 million
- Annual cost of triple combination therapy in US – US$10,000–15,000
- Annual cost of triple combination offered by a generic Indian manufacturer (quality meeting international standards) – US$800–1000

Source: 'Six Month Report Card: have AIDS drugs prices for the poor been slashed?', Médicins sans Frontières, 1 December 2000, accessed at: http://www.accessmed-msf.org/msf/.

Many NGOs were sceptical of the initiative at the beginning, and their concerns have been validated. Seven months after the scheme was launched, Médecins Sans Frontières (MSF) published a report card. The findings were staggering. Instead of enjoying across the board price reductions, poor countries have to negotiate individually with each company for each drug. This both weakens their negotiating position and consumes huge human resources. After a year, few benefits had ensued.

The May 2000 initiative represents one of the clearest examples of the pharmaceutical companies' role in influencing the health governance agenda. It has deflected attention from the development of more sustainable solutions, regionally, nationally or locally. Even within its own limited vision, this initiative has failed to deliver. What is more, this initiative was ongoing while 42 companies, including some of these five, were still fighting the Medicines Act in the High Court in South Africa. The case was tied up there for three years and during this entire time people were dying due to lack of availability of affordable drugs. As mentioned earlier, even if the price of the drugs were to plummet, big investments would be needed to develop the necessary infrastructure for their appropriate delivery. This lack of infrastructure, however, in no way legitimates the continuation of drug prices beyond the means of governments and people.

11.7   Looking forward

The focus on the case of access to ARV drugs highlights the problematic role of pricing and patent right protection in devising responses both at the global as well as the national levels to the AIDS epidemic. In practice the legal provisions under the WTO for compulsory licensing and parallel imports, rather than providing a guarantee for genuine public policy responses in the case of emergencies, instead seem to be providing transnational corporations with an opportunity to contest such policy responses. The prime objective of US trade policy – followed and supported by the EU Trade Commission – and transnational pharmaceutical companies has been to assert, in principle, the primacy of TRIPS, and thus secure projected profits: the case of ARVs as elucidated in this chapter illustrates this.

This stance is sustained by arguing that it is not patents, or the price of ARVs, which are the problem; that many drugs needed in the South are not subject to patent; and that in any case many of the poorest Southern states do not have to comply with the TRIPs until 2006. Access problems are presented from this perspective as mainly due to infrastructural failings, security issues (military conflicts), and lack of expertise in the concerned Southern states (*Health Horizons*, 2000).

At the broadest level, the issue of access to ARVs underscores the struggle at the global level between two competing political projects. On the one hand, there is the neoliberal project, concerned with firstly disembedding the market from political influence, and secondly expanding its reach across social institutions. On the other hand, there is a social-democratic project concerned with the delivery of welfare provisions on a more egalitarian basis rooted in conceptions of social justice. As global protests and increased campaigning in the light of the heavy-handed approach taken by transnational pharmaceutical companies and the US government began to make the neoliberal stance less acceptable, attempts at a diffusion of that stance were under way. It is in this context that the latest proposals for public–private partnerships have to be understood and evaluated.

As we have seen, the particular public–private initiative outlined in this chapter is essentially flawed, and reveals some major shortcomings, which may turn out to be endemic to the new public–private approach to public health. It raises, for instance, the question of whether it is possible for a true partnership to develop in a situation of structural inequality? What are the prospects for an effective partnership between transnational companies and under-resourced UN agencies, or between multilateral organizations and developing countries?

It is clear that next to the issue of pricing, co-ordinated efforts for managing outreach, distribution and effective administration of drugs have to be in place, efforts which involve a much enhanced role for both the institutions of the UN co-ordinating globally and local agencies, groups and institutions. Yet, the current situation reflects the stance, voiced by one African spokesperson, that '... there are discussions about us, excluding us' (Tshabalal-Msimang, 2000). Robin Stott points out that for partnerships to be effective, health partners need to make policy and hold budgets together (Stott, 1999, p. 822). In addition, necessary ingredients for success include drugs access that is broad (in terms of range of therapeutic remedies), inclusive (reaching all who need them) and sustainable (encouraging regional or national self-sufficiency).

What role is there for global governance to make a positive impact on behalf of AIDS sufferers and the states concerned with their welfare, and where could such input originate? The WHO, relatively sidelined as a multilateral agency in the 1990s, is now taking a more active interest in trade policy and may make an important contribution to the current policy debate. In May 1999, the 52nd World Health Assembly gave WHO the mandate to do more work on trade-related issues, including access to drugs. In particular, WHO was asked to study the effects of international trade agreements on health. NGOs are working with the WHO to track prices and access to essential drugs (Duckett, 1999, p. 7). Governments need this data if they are to comprehensively increase their chances of providing access to AIDS-related medication.

Furthermore, the WHO has a role in helping developing countries understand the health implications of the WTO and the TRIPS. WHO/DAP produced 'Globalization and Access to Drugs: Perspectives on the WTO/TRIPS agreement', to help with interpretation and guidance. SADC countries have called on WHO to participate in WTO negotiations and help draft national laws safeguarding compulsory licensing and parallel importing (Kaiser Daily HIV/AIDS Report, 2000b). Potentially, UNAIDS could help here as well. Clearly, it would be helpful if model legislation could be drawn up for developing countries to help them in the development of national intellectual property laws.

Support for a strengthening of the public policy role of UN institutions *vis à vis* the imbalances created through the privileging of investors' interests through the WTO is beginning to build. While states like South Africa, Thailand, Brazil and India have acted individually, we are beginning to see strength in numbers and more concerted action not only in the domestic arena, but internationally. This was clear, for example, at the World Summit for Social Development held in Geneva in June 2000, when the G77 countries pushed a proposal for the final conference text to exclude essential medicines from patentability. While they did not achieve this, at least they did succeed in getting an affirmation of countries' rights to freely exercise their legal options (Oh, 2000).

Furthermore, health activists and scholars are also beginning to realize and engage the potential significance of the WTO, not only for general issues of inequality, but for health issues in particular (Baris and McLeod, 2000; Labonte, 1998). Awareness raising and campaigning on health-trade issues is gaining momentum, as activists see the importance of putting health at the centre of trade debates, rather than on the periphery. The example of US–Brazil–WTO cited earlier illustrates how rapidly NGOs can organize a response.

Access to drugs is affected deeply by a number of factors, one of which is clearly price. The question is whether the governance framework for public health policy will continue to favour an individual's ability to purchase expensive patented drugs over broad-based access and the expanded use of cheaper generic products. Global health issues cannot be abstracted out of *public* policy. If this is so, then a response to health insecurities framed in terms of the former option can only be an *inappropriate* political response. If the global social policymakers are sincere in their commitment to seek solutions to the global health crisis, *there is no alternative* but to radically rewrite the regulations of the global health governance agenda so as to situate access to health care within the parameters of a true public good: the politically guaranteed provision of health care to all based on *need* rather than primarily ability to pay.

Acknowledgement

This chapter was originally published in *Third World Quarterly* vol. 23(2), April 2002.

12
Global Health Governance: Some Theoretical Considerations on the New Political Space

Ilona Kickbusch

12.1 A new political space and paradox

It is difficult enough to keep up with the rapid transformations in global health, let alone provide a robust theoretical framework to analyse the developments underway. This is further complicated by the gulf that divides scholars of policy/International Relations and public health. As a result, explanatory models that political science may provide for the shift in international health and its future directions remain as elusive to the public health debate, as global health remains a non-issue for scholars of global governance. Some inroads have been made as the public health community tries to gain a better understanding of the process of globalization (Kickbusch and Buse, 2000; Lee, 2000), but there remains a significant lack of scholarly work that provides a systematic analysis of global health governance.

In contrast, in the political realm the global health debate is increasingly being conducted far beyond its usual confines. Global health issues have gained a new dimension, a new immediacy and a human face as Africa's AIDS crisis is transported by the global media: 'AIDS is the number one issue in the world today, the number one issue. The level of the AIDS crisis, its potential to destroy economic achievement, undermine social stability and create more political uncertainty...is enormous.' This quote by the then United States ambassador to the United Nations (UN) Richard Holbrooke, on occasion of the historic UN Security Council debate on HIV/AIDS in March 2001, expresses the shift in perspective and opens up an interesting *political paradox*. While the predominant reference frame is narrow and focused on the fight against infectious disease, the political response has broadened. The protection of health is no longer seen as primarily a humanitarian and technical issue relegated to a specialized UN agency, but more fully considered in relation to the economic, political and security consequences for the complex post-Cold War system of interdependence. This has led to new policy and funding initiatives at many levels of governance and a new political space within which global health action is conducted.

In January 2001 the response to the global infectious disease threat was a significant part of the World Economic Forum; it also featured prominently in European Union meetings and bilateral summits, as between French President Jacques Chirac and US President Bill Clinton in early 2001. The March 2001 UN Security Council debate on HIV/AIDS followed on the G8 deliberations at the Okinawa Summit in July 2000, at which the leaders of the major industrialized countries committed themselves to halve the global infectious disease burden by 50 per cent by 2010. The G8 summit in Genoa in July 2001 further explored the mechanisms to achieve this goal, building on the outcome of the United Nations General Assembly Special Session (UNGASS) on HIV/AIDS in June 2001. In preparation for UNGASS the UN Secretary General made an unprecedented historical move in proposing a new global financing mechanism in the form of a Global Fund to fight AIDS, Tuberculosis and Malaria, a proposition that he has taken to world leaders, to the business world and to the member states represented at the May 2001 World Health Assembly. US President Bush made US$200 million available as the first major payment into such a fund, other developed countries followed at the Genoa summit and the first group of developing countries has now also contributed to this fund (Brugha and Walt, 2001; Office of the Spokesperson for the Secretary General, 2001).

Broader global health issues feature prominently in international financial and economic institutions, as in the controversial trade and intellectual property rulings of the World Trade Organization with regard to patent rights on drugs and the negative health impacts of structural adjustment programmes. Health is one of the largest global growth industries with extraordinary profit margins, particularly in pharmaceuticals. Health has become one of the most tangible areas through which NGOs can demonstrate extreme global inequalities and negative impacts of globalization, as witnessed through demonstrations on occasion of the key meetings of the World Bank and the World Trade Organization in 2000 and 2001. Health has also moved back to the forefront of development debates and discussions of poverty eradication. For example, the WHO Commission on Macroeconomics and Health established by Director-General Gro Harlem Brundtland is arguing for significantly increased health investment in the poorest countries in order to ensure economic and social development. The report of the Commission was completed in December 2001. And health is a key challenge for the existing member states of the European Union, as they seek to more closely harmonize their public health regulations and as they consider enlarging memberships to include countries with significantly lower life expectancies and struggling health systems.

12.2　The search for an analytical framework

Why is this happening? What is driving this flurry of political energy? How is it related to globalization? Who is in charge? How does global health differ from international health? How will it change the face of health and health policy as we know it? One way of answering such questions systematically is to draw on International Relations theory and the growing literature on global governance. While there is no universally accepted definition of global governance the increasing use of the term in itself reflects the recognition that 'international relations' as a system is no longer sufficient to describe the very diverse formal and informal sets of arrangements in global politics (Makinda, 2000).

In both political theory and by extension political practice, several schools of thought – realist, liberal and constructivist – provide competing theoretical frameworks and interpretations. Makinda (2000) provides a useful short overview of the theoretical frameworks used to explain and understand global governance. While there are many facets of (and increasing overlaps between) these three schools, they can be summarized as follows. *Realist* approaches are state-centric and emphasize that nation states use the multilateral system to maximize their national interest. The *liberal* perspective takes the agenda beyond security and self-interest to include international law and a human rights agenda. It includes the many other actors in the international arena and believes that international organizations and the UN in particular play an important role in world politics by helping to establish and maintain some order within anarchy. The *constructivist* approach focuses on the norms, rules and social institutions that make up the global system and constitute the identities and interests of states and other international actors, and enable it to learn and go beyond self-interest towards a global agenda and a global system.

Applied to global health governance, the realist school would focus on the co-operation of sovereign states in global health as interdependent utility functions based on self-interest, security considerations and competition for power; the liberal frameworks would expand the focus beyond state actors and contend that the commitment of nation states to global health is also driven by collective considerations such as human rights and individual freedom, particularly with a view to the resolution of market failures; and a constructivist approach would look at the system of global health as a whole and analyse the ideas and motivations that drive and shape the system. America's interpretation of infectious disease as a national security priority is clearly a realist strategy, present Scandinavian and British health development policies tend to follow a more liberal framework committed to closing the health gap and ensuring global public goods, and both the UN system and many of the new actors in the health arena would look to establish new mechanisms of norm setting and brokerage of competing interests in a social constructivist mode.

Does it matter according to which theoretical framework we interpret the world? Does it matter what academics think or what policymakers think? I would argue – with Thomas Kuhn (1996) – that the predominant intellectual framework and rationale significantly influences the practice and priorities of global public health, in particular the accepted architecture of governance, the membership, the responsibilities of the individual members and the long-term commitment to a normative order that drives global health governance (Kickbusch, 2000). Epistemic communities – transnational networks of policy professionals who share common values and causal understandings (Haas, 2001) – exert significant influence on how issues are perceived, how problems are framed and what solutions are proposed. The framing of HIV/AIDS as a human rights issue is a case in point, as is the global burden of disease measurement or was the flurry of 'health care reforms' in the 1990s.

One of the most extraordinary attempts to frame a global health agenda, the Health For All (HFA) movement launched by WHO in the late 1970s, has not to this day received the analytical attention it deserves, leading to much reinvention of the wheel as political agendas move in cycles.

What, for example, is the difference between the HFA framework and goal of 'a level of health that will permit all people to live a socially and economically productive life' and the approaches proposed by the Commission on Macroeconomics and Health? Will the framing of the global health agenda as a 'war' against infectious diseases lead to a higher level of commitment? Will it provide inroads to the broader health agenda and the determinants of health? How will the political paradox play out? More detailed analysis can provide us with better insights as to how countries use the UN to promote their own health agenda, or use health to promote other agendas; it can show how global priorities are established, which power disparities exist and whose values are promoted. Most importantly, analysis can provide transparency and contribute to the accountability of a global health system and its leaders, and in the long run, it can support legitimacy by dispelling myths. Different insights and interpretations by different schools of thought can only help to clarify the issues at stake. Above all, theory defines the ontological basis for how global health is perceived – its territory and its logic – which in turn impacts on the priorities set in the political arena.

12.3   A social constructivist agenda

Of the three schools of thought described above – and there have been other groupings and classifications used in the study of international relations as is usual in any lively discipline – it is my view that a social constructivist framework (Ruggie, 1998) offers the best theoretical starting point to help understand the dynamics of global health governance and the role of the UN system. But we must also consider the realist and liberal frameworks in order to understand the rationales of individual actors, in particular nation states, within the global governance system. Guidance also comes from the very rich literature on environmental governance. In his analysis of 30 years of international environmental governance, Peter Haas (2001) concludes that what he terms the transformative 'dynamic' view has been superior in accounting for changes over time in environmental governance: 'it has been better able to account for the mechanisms by which states' notions of the national environmental interest have changed as a consequence of their involvement in international environmental regimes, and their exposure to international institutions'. Ruggie (1998), probably the leading constructivist thinker who recently served for several years as adviser to the UN Secretary General, provides us with the best introduction to the *social constructivist approach*. He maintains that a new type of 'collective intentionality' is developing in the international system, which reaches beyond competing and self-interested nation states and is moving towards a new value base, rationale and practice of international co-operation. He proposes to regard international relations (and by extension global governance) as a social framework and perceives international institutions as 'systems of learning' that create new rights and responsibilities for its members through a process of 'collective legitimization'. The constructivist approach insists that ideas do make a difference and that aspirations, legitimacy and rights can drive global action. In doing so it explicitly builds on Giddens's (1981) notion of agency and Habermas's theory of communicative action (Habermas, 1984, 1987).

I would suggest that in health we have lost the 'collective intentionality' for a broad health agenda that was successfully established through the leadership of WHO in the Health For All movement and Primary Health Care strategy. What we are experiencing instead is an intensified process in the political arena of establishing a 'collective intentionality' for addressing HIV/AIDS. The development of perceptions and policies around HIV/AIDS provide an excellent illustration of social constructivism as an analytical tool (Kickbusch, 2001):

- a value base is established (health as a human right, AIDS as a human rights issue),
- collective learning is organized through international organizations, NGOs and the global media (AIDS as a global threat and a common responsibility),
- new responsibilities of the various actors evolve (in particular with respect to political leaders, the pharmaceutical industry and the World Trade Organization),
- new rights are established (with regard to access to essential medicines as a right over trade),
- new rationales evolve (health as a security and foreign policy issue, health as a global public good, AIDS care as an inroad to build health systems),

- new actors gain legitimacy (the role of NGOs, the new philanthropists),
- new practices, mechanisms and institutions evolve (the proposed UN Global Fund to fight AIDS, Tuberculosis and Malaria, the International Aids Vaccine Alliance, the Global Health Access Project).

HIV/AIDS also allows for a comparison between realist, liberal and social constructivist policy approaches. Namely, is the response based on self-interest and threat, on responsibility of the global community to address market failures and establish global public goods, or does it provide the basis for a new approach in global problem solving with significant new collective legitimization for non-state actors? Obviously all of these perspectives interact in the global arena to create the dynamic new political space that HIV/AIDS represents at this point in history. Whether this space will allow for the introduction of a broader global health agenda and a 'collective intentionality' remains to be seen. Will it address root causes and the construction of health infrastructures or will it lead to a focus on one disease at the expense of addressing health as a whole? This ambiguity remains the key challenge for all global health activists.

12.4   The new primacy of politics

So what of the forces of globalization? Zakaria (2000) has recently argued that the dominance of economics that followed the end of the Cold War led nation states (and many other actors) to underestimate the new quality of politics necessary to respond to globalization. This 'new quality' has several dimensions. The end of the bipolar world and the global advance of democracy are part of the solution and part of the problem. While many countries now share an interest in advancing democratic principles, they are not bound by Cold War ties: 'no one has said we must vote for each other when interests clash' (Crossette, 2001). The increasing number and diversity of players, the visibility through global media and the pressure towards transparency and accountability from global civil society calls for new skills in international negotiations that far transcend traditional diplomacy and traditional public health. Trust continually has to be built within a fluid 'network of governance' (Reinicke, 1998), where alliances are renegotiated constantly. So while sovereign states presently remain the agreed dominant political framework – as at the meetings of the World Health Assembly (WHA) – it still follows that constructing or analysing global health from a state-centric perspective bypasses one of the most significant developments in the global health arena: the creation of a new political space or, to quote Chris Patten (2000), 'a new political ecosystem'.

A key challenge we face is to analyse how the evolving dynamics of this political ecosystem are being shaped by the driving forces of globalization and in particular by the actions of the new kids on the block: the health activists and NGOs, the new global philanthropists and the private sector, as well as the dense network structure they have created. Their interaction with the existing institutions of global governance, with nation states and with one another (most frequently in the form of new and overlapping alliances) is changing the global playing field in health, its norms, its rules, its practices and above all its power politics. They are hungry for resources or political attention, or both, and are engaged in a constant back and forth between coalition and competition. Increasingly these players are demanding to be recognized as constitutive units of the global health system and are developing (as is appropriate for an ecosystem) a wide variety of organizational forms, which for lack of a better term are usually grouped under the umbrella of 'public private partnerships' (Wolfgang *et al.*, 2000) or 'global public policy networks' (Buse and Walt, 2000a).

In a few years the Bill and Melinda Gates Foundation has pump-primed global health development and demonstrated that money does make a difference and sets agendas overnight. Health activists have been able to shift the political and moral debate in global health and make it no longer acceptable that 95 per cent of the 34 million people living with HIV/AIDS have no access to triple drug therapy. The public pressure on pharmaceutical companies has led to new pricing mechanisms as well as large donations that significantly reduce the costs of drugs to fight a range of infectious diseases in developing countries. It is, as Giddens (1999) has highlighted in his Reith lectures on globalization, not only increasing interdependence and speed that drives the global system, but a new non-territorial political space and a qualitative shift of problems and solutions. Paquet (2001) analyses how the geo-political space is constantly shifting, introducing a 'network paradigm' that understands the qualitative change in the patterns of powersharing between polity, market economy and civil society and between levels of governance, which necessitate constant learning and adjustment between economy, state and society and between various levels of subsidiarity. In this new political space the international organizations are only one of the many players, frequently not very rich and not very powerful and not as steeped in legitimacy as in the past, albeit with exclusive access to sovereign nation states through their governing bodies. But, social constructivists would argue, they have unique opportunities to act as brokers, organize global learning, set norms and standards and, in some cases, move into the area of regulation. Indeed, as Kofi Annan has demonstrated through the Global AIDS and Health Fund, they have enormous potential for leadership and agenda setting. As collaboration becomes the new categorical imperative, 'strategic organizations' with collaborative capabilities gain in importance. They engage through a mixed process of coherence and pluralism in a world where the 'center … is more a network than a place' (Handy, 1995).

12.5   What role for the realist American hegemony?

An additional complexity in the global health arena is the position of the United States, which falls squarely into a realist framework of 'interdependency utility', a position it rarely abandons in the global arena. At its root lies America's combined sense of exceptionalism and vulnerability. Exemplary is the argument put forward five years ago in a report by the US Institute of Medicine (1997), *America's Vital Interest in Global Health*, stating that it is both in the political and the economic interest of the US to invest in combating disease abroad. Other reports have expanded this argument but stayed firmly within the realist framework, including The National Intelligence Estimate (Gannon, 2000) and the report on *Health as a Global Security Challenge* issued by the Chemical and Biological Arms Control Institute together with the CSIS International Security Program (2000). A recent initiative by the Foreign Affairs Council (Kassalow, 2001) defines health as a key issue of foreign relations. 'A foreign policy that gives higher priority to international health is good for the United States and good for the world', the report concluded on the eve of Secretary of State Colin Powell's first trip to Africa, which gave the AIDS issue a high policy profile.

The reports underline the need to include health in both the security and the foreign relations agenda. Because it is no easy task to convince these constituencies to show concern for these 'soft' issues, however, they stay very firmly rooted in the framework of national sovereignty and US hegemony, and offer little on improved mechanisms of global health governance. A number of political and historical reasons can explain why the US in particular has an affinity to realist and utilitarian rationales (Chase, 1988; Ruggie, 1998), and the American public health community has been very astute in arguing within the preferred framework of 'threat' rather than interdependence. Based on this rationale the Clinton administration made the first ever appointment of a health adviser to the National Security Council, and the US Congress significantly increased the funding for health research and US foreign assistance. On the whole, the public health community welcomed the higher profile and funding for public health; only recently have a group of authors (Fee and Brown, 2001; Geiger, 2001; Henretig, 2001; Sidel *et al.*, 2001; Wetter *et al.*, 2001) sounded a word of warning about subjecting the public health ethic to one of security and military mindframes.

What are the implications for global health development if the most powerful state continues to be driven by a realist frame of thought and action? On several occasions, health has been a test case for unilateral approaches rather than multilateral consensus; for example, 20 years ago in relation to the Code for the marketing of breast milk substitutes and most recently in relation to patent laws for medicines, where on both counts the United States chose a protectionist pro-industry approach rather than join a global effort on health protection. There are indications following the recent US moves to not join a whole number of international treaties that the US position will also change with regard to support of the framework Convention on Tobacco Control. 'Over the last decade, as the United States looked down from the peak of world power, its officials may have over-looked some currents rising below', writes the UN correspondent of the *New York Times* in commenting on the exclusion of the United States from the UN Human Rights Commission (Crossette, 2001). Political commentary is increasing in the US, which warns against unilateralism and proposes that in order to move productively in the new global political space, the US will need to reassess both its role and its approaches and learn to accept interdependence and international institutionalization. In the long run, say some analysts, the United States 'will be in a stronger position to lead if it supports the United Nations' (Russett and Oneal, 2001). Whether this happens or not will have a crucial influence on the development of global health, its direction, its priorities and its funding.

12.6   A power vacuum or a new space?

We are in transition from what seemed a relatively reliable, state defined and structured world of international health to a diffuse political space of global health in which new forms of distributed power and new patterns of power-sharing emerge (Paquet, 2001). We need to analyse to what extent the political ecosystem that inhabits this space transfers power and to whom. We need to map the epistemic communities and the multitude of networks and their spheres of influence. In a world without a central authority leadership, accountability and transparency take centre stage. Increasingly the developing world is watching with scepticism how 'global priorities' become just another linguistic expression of the interest of the rich and powerful countries, and the plight of the poor is not improved, despite increasing globalist rhetoric. The increasing tension between the US realist and the European liberal approaches will also play out in the health arena, as will the conflicts between the developed and developing nations over financial contributions to the UN organizations. For example, the budget compromise reached at the UN in order to appease the US Congress and induce it to pay its arrears to the UN has led to a US$20 million budget shortfall for the WHO, as the group of 77 countries in turn refused to increase their contributions in line with the resolution adopted by the UN General Assembly. Are we then, as Orbinski (2001) contends, in a global governance vacuum that breeds the political and financial disengagement of nation states from international institutions and leaves open a space for the global market, or is it that we have entered a different game with new rules in a new political space?

A case in point that deserves detailed analysis is the turnaround of the pharmaceutical industry in just a few months with regard to patent laws and pricing of HIV/AIDS drugs for lower-income countries, which, despite Orbinski's postulate of a power vacuum, was brought about not least by organizations like his own, Médicins Sans Frontières (MSF). Again the experience with new forms of global governance and regime formation from the environmental governance research could prove helpful in disentangling the issues at stake (Young, 1997). The disturbing fact remains that the 'old' system of international relations did not live up to its responsibilities, and even the most cursory analysis of global health governance shows that existing mechanisms of governance and resource flows are insufficient to address the problems at stake and to generate the political will necessary for action.

Nowhere has this been as obvious as in the failure to react forcefully to the global threat of HIV/AIDS, and the chips are still down on whether the new global impetus will constitute a real turning point in health governance. Laurie Garrett (2000b), in her analysis of global public health, speaks of a 'betrayal of trust' of an unparalleled global dimension, very reminiscent of a March of Folly as described by Tuchman (1984). A world that had the solutions at hand was not willing to apply them and to share its resources with the poorest. While the political awareness and commitment seems to be growing as described above, the financial gap to support access to health and health care in the poorest countries remains staggering. Jeffrey Sachs, chairman of the WHO Commission on Macroeconomics and Health, has stated bluntly: 'what developing countries need to successfully fight infectious disease are three things: money, money and money' (statement at ECOSOC, April 2001). The Global Fund on AIDS and Health is calculated at US$10 billion; establishing access to basic health care in poor African countries is estimated to cost US$60 per head.

12.7   Constructing a virtuous governance cycle

But is Jeffrey Sachs right? Is more money really the key ingredient to a solution? Increasingly analysts warn that we must re-establish the primacy of politics and agency rather than continue to discuss globalization as if it were a force of nature (Messner, 1999). Zakaria (2000) reminds us in his short but succinct analysis that the last era of globalization in the late nineteenth century was undone not by bad economics but by bad politics, just as the AIDS crisis was not brought on by lack of health and medical expertise but by the infuriating unwillingness of political leaders to face and accept the issue and take action. Applied to the global health debate this means that money is crucial, but that agency and politics are just as essential. Health needs the commitment of the political leaders of sovereign nation states to establish reliable health infrastructures, and they in turn need the support of the global community to do so.

Building the elements of a global governance architecture is a key challenge for the early twenty-first century, but (using an analogy from architecture) it will need to look more like a Frank Gehry structure than a Le Corbusier building. And we do not have much time to build it. In the face of the AIDS crisis the world community 'needs to come up with a strategy that is well governed and moves money quickly. You've got to keep your eye on the prize, not on the process' (Bernard, quoted in Kassalow, 2001). Yet good health governance implies an inclusive process in order to be sustainable in the long run, otherwise we will continue to chase disease after disease in one health war after the other. 'We need not just effective governance, but the right kind of governance' is the guidance given by Robert Keohane (2001) in a recent presidential address on governance and globalization in which he underlines the responsibility of political science to contribute to the design of institutions for a globalizing world. He suggests that such design must bring together ideals and reality. One attractive element of a constructivist view is that it allows for learning, not just of individual states but of systems, a view compatible with a more dynamic understanding of reality and with the complexity of the global health arena. But learning also means looking back and developing a better understanding of why the world failed to follow through on its Health For All commitment and why political leaders allowed global health to deteriorate as it did. Laurie Garrett's work (2000b) on the demise of global public health needs to be followed up with in-depth studies that allow us to understand why, in the course of globalization, the primacy of economics overran all knowledge of what constitutes sound health policy. Former Finnish Director General of Health Kimmo Leppo has frequently drawn our attention to this feature of health policy: 'One of the great paradoxes in the history of health policy is that, despite all the evidence and understanding that has accrued about determinants of health and the means available to tackle them, the national and international policy arenas are filled with something quite different' (Leppo, 1997).

Global health governance arrangements depend on committed human effort. Bruce Russett and John Oneall (2001) have proposed that there can be virtuous cycles in world politics, which are constituted by the interplay of democracy, economic interdependence, and international law and institutions, a concept of peace originally developed by Immanuel Kant. Their statistical analysis shows that the density of international ties reduces conflict and, in combination with economic interdependence and democracy, generates a range of positive systemic effects. The major conclusion of their study is that 'there is a basis for a dynamic international system that is able to perpetuate and enhance itself'. It seems, therefore, that present developments in the global health arena, with its ever-increasing number of actors and ever-denser networks of persuasive rather than coercive nature, are steps in the right direction and need to be reinforced rather than seen as chaotic aberrations. But just as important are new mechanisms of interstate co-operation. Because national sovereignty will continue to be with us for some time to come, the pooling of sovereignty in key health matters needs to be explored (Reinicke, 1998), as is beginning to be practiced in the European Union and in proposals for new global conventions, such as the proposed Framework Convention on Tobacco Control (FCTC). At the other end of the spectrum, mechanisms for 'supportive health intervention' in failed states need to be considered that go beyond present approaches to humanitarian assistance. New financing mechanisms are of prime importance and the establishment of a new structure such as the Global Fund to fight AIDS, Tuberculosis and Malaria must be seen as the beginning of the attempt to create a new 'collective intentionality' for broader global health goals and new global health governance mechanisms.

The challenge before us is to identify the key elements of what would constitute a virtuous cycle of health governance and to explore these components with creative and rigorous research. Following Kant's lead, I propose it would be the interface between:

- equitable and sustainable national health systems
- strong international institutions and dense international health networks of state and non-state actors committed to transparency and accountability, and
- a respected system of international health regimes, laws and regulations.

A sound and clear theoretical framework will help in the design process of global health governance and in the ensuing debate on the normative framework that accompanies any institution building. But it is politics and political commitment that must continuously counteract the forces of fragmentation and pure self-interest. The point, therefore, is not to answer the question 'Who is in charge?', but how political leadership should be exercised in the global arena; to paraphrase a quote by John Ruggie (1998) 'Creative leadership is social constructivism in action'.

References

Achmat Z., 1999, 'We Can use Compulsory Licensing and Parallel Imports: A South African Case Study', Aids Law Project, South Africa, November/December.

ACT UP Paris, 2000, posted on INTAIDS [intaids@hivnet.ch] 26 February, originally published in *Le Monde*, 29 January 2000.

African Development Forum, 2000, 'Aids Consensus and Plan' December, on the Economic Commission for Africa website at http://www.uneca.org/adf2000.

Ahn D., 2000, 'Linkages between international financial and trade institutions. IMF, World Bank and WTO', *Journal of World Trade*, 34(4): 3–34.

Ainsworth M. and M. Over, 1992, *The Economic Impact of AIDS: Shocks, Responses, and Outcomes*, World Bank Technical Working Paper 1, Africa Technical Department (Washington DC: World Bank).

Ainsworth M., L. Fransen and M. Over, 1997, *Confronting AIDS: Public Priorities in a Global Epidemic* (New York: Oxford University Press).

Aitken D., 1994, Letter to Mr K. Ramanth, ITC Ltd, Re: Aitken's visit to India, Guildford Depository, 23 February, Bates No.: 301710327-329.

Altenstetter C. and J. Bjorkman, eds, 1997, *Health Policy Reform, National Variations and Globalization* (London: Macmillan).

Altman D., 1999a, 'AIDS and questions of global governance', *Pacifica Review* 11: 195–211.

Altman D., 1999b, 'Globalization, political economy and HIV/AIDS', *Theory and Society*, 28: 559–84.

Altman D., 2001, *Global Sex* (Chicago: University of Chicago Press).

Alubo S.O., 1990, 'Debt crisis, health and health services in Africa', *Social Science and Medicine*, 31(6): 639–48.

Amoore L., R. Dodgson, B. Gills, P. Langley, D. Marshall and I. Watson, 1997, 'Overturning globalisation: resisting the teleological, reclaiming the political', *New Political Economy*, 2(1): 179–95.

Andrews E., 1997, 'WTO overrules Europe's ban on U.S. hormone-treated beef', *New York Times*, 9 May.

Annan K., 2000, 'Macroeconomic policy questions: external debt crisis and development. Recent developments in the debt situation of developing countries' Report of the Secretary General, UN A/55/422 General Assembly 26 September, at http://www.un.org/documents/ga/docs/55/a55422.pdf.

Anon. Professional Developments, 1996, 'OECD Symposium on the Future of Public Services', *Public Administration and Development*, 16: 281–5.

APUA, 2001, 'Alliance for the Prudent Use of Antibiotics', at http://www.healthsci.tufts.edu/apua/apua.html.

Armstrong D., 1982, *The Rise of International Organisation: A Short History* (London: Macmillan), p. 43.

Asia Link Consulting Group, 1991, 'Qualitative research for 555 State Express Among Chinese Americans', Guildford Depository, July, Bates No.: 503852931-2958.

Asian Harm Reduction Network, 1998, 'The Hidden Epidemic', *Asian Harm Reduction Network Newsletter*, No. 10 January/February.

Asthana S., 1994, 'Economic crisis, adjustment and the impact on health', in D. Philips and Y. Verhasselt, eds, *Health and Development* (London: Routledge).

Bailey K.V. and A.H. Ferro-Luzzi, 1995, 'Use of body mass index of adults in assessing individual and community nutritional status', *Bulletin of the World Health Organization*, VXXIII(5): 673–80.

Baris E. and K. McLeod, 2000, 'Globalization and international trade in the twenty-first century: opportunities for and threats to the health sector in the South', *International Journal of Health Services*, 30(1): 187–210.

Barks-Ruggles E., 2001, 'Meeting the Global Challenge of HIV/AIDS', Policy Brief No. 75, April, Brookings Institution http://www.brook.edu/comm/policybreifs/policybriefs/pb075/pb75.htm.

Barnes E., P. Hanauer, J. Slade, L.A. Bero and S.A. Glantz, 1995, 'Environmental tobacco smoke: the Brown and Williamson documents', *Journal of the American Medical Association*, 274: 248–53.

Barnett T. and A. Whiteside, 1999, 'HIV/AIDS and development: Case studies and a conceptual framework', *European Journal of Development Research*, 11: 200–34.

Barua D., 1972, 'The global epidemiology of cholera in recent years', *Proceedings of the Royal Society of Medicine*, 65: 423–32.

Bassett M. and M. Mhloyi, 1991, 'Women and AIDS in Zimbabwe: The making of an epidemic', *International Journal of Health Services*, 21: 143–56.

Bassett M.T., L. Bijlmakers and D. Sanders, 1997, 'Professionalism, patient satisfaction and quality of health care: experience during Zimbabwe's structural adjustment programme', *Social Science and Medicine*, 45(12): 1845–52.

BAT, 1983, 'BAT Board Guidelines: Smoking Issues: Part II General Strategies', Guildford Depository, March, Bates No.: 201766240-246.

BAT, 1991, 'European Tobacco Advertising – Protecting BAT's interests', Guildford Depository, 17 January, Bates No.: 202028839-8840.

BAT, 1995, 'Key Area Paper: Indirect Taxation of Tobacco Products', Guildford Depository, 18 April, Bates No.: 503900886–0895.

BAT, 2000a, 'BAT Annual report, 1999', http://www.bat.com/annualreport/reviewc.htm (accessed 13 June 2000).

BAT, 2000b, 'News release: British American Tobacco Proposes "Quantum Leap" for Sensible Tobacco Regulation', 29 August http://www.bat.com (accessed 11 September 2001).

BAT, 2000c, 'News Release: Tobacco advertising bill', 6 December http://www.bat.com (accessed 11 August 2001).

BAT, 2001a, 'Annual Review and Summary Financial Statement 2000' http://www.bat.com (accessed 9 August 2001).

BAT, 2001b, 'What We Do: International Brands' http://www.bat.com (accessed 9 August 2001).

BATCo, 1988, 'World Health Assembly – Geneva May 2–14 1988, Contact Programme', Guildford Depository, Bates No.: 502586873-6877, cited: Committee of Experts (2000).

BATCo, 1993, 'Lucky Strike Sponsorship', Guildford Depository, Bates No.: 303593886-3900 http://www.library.ucsf.edu/tobacco/batco/html/500/560 (accessed: 11 August 2001).

BATCo Marketing Intelligence Department, 1994, 'International Brands 1988–1992', Guildford Depository, January, Bates No.: 500056134-6179.

BAT Indonesia, undated, 'A Study on the Smokers of International Brands', Guildford Depository, Bates No.: 400458935-9056.

BAT Marketing Department, 1984, 'Trademark Diversification Activities of Selected International Brands', Guildford Depository, September, Bates No.: 303511934-975.

BATUKE, 1992, 'BATUKE response to BATCo Draft Policy, Health Warnings on Cigarette Packings', 24 April 1992, Bates No.: 301643934-3935.

Baum F., 2001, 'Health, equity, justice and globalisation: some lessons from the People's Health Assembly', *Journal of Epidemiology and Community Health*, 55(9): 613–14.

Baylis J. and S. Smith, eds, 1997, *The Globalization of World Politics, An introduction to international relations* (Oxford: Oxford University Press).

BBC News, 1999, 'Tobacco giant admits health risks', 13 October http://news.bbc.co.uk/hi/english/health/newsid_473000/473630.stm (accessed 11 August 2001).

Beaglehole R. and R. Bonita, 1997, *Public Health at the Crossroads, Achievements and Prospects* (London: Cambridge University Press).

Beaglehole R. and R. Bonita, 1998, 'Public health at the crossroads: which way forward?', *The Lancet*, 351: 590–2.

Bennett R. and D. Blackwell, 2001, 'Tobacco Smuggling: Inflation-only cigarette tax rise may be repeated', *Financial Times*, 9 March.

Bernard K., quoted in J.S. Kassalow, 2001, *Why Health Is Important to U.S. Foreign Policy* (New York: Council On Foreign Relations & Milbank Memorial Fund).

Bero L., E. Barnes, P. Hanauer, J. Slade and S. Glantz, 1995, 'Lawyer control of internal scientific research: the Brown and Williamson documents', *Journal of the American Medical Association*, 274: 241–7.

Bertozzi S.M., 1996, 'The impact of human immunodeficiency virus/AIDS', *Journal of Infectious Diseases*, 174(Supplement 2): S253–7.

Bettcher D. and I. Shapiro, 2001, 'Tobacco control in an era of trade liberalisation', *Tobacco Control*, 10: 65–7.

Bettcher D., S. Sapirie and E. Goon, 1998, 'Essential public health functions: results of the International Delphi Study', *World Health Statistics Quarterly*, 51: 44–54.

Bettcher D. and D. Yach, 1998, 'The globalisation of public health ethics?', *Millennium: Journal of International Studies*, 27(3): 495.

Bettcher D.W., D. Yach and G.E. Guindon, 2000, 'Global trade and health: key linkages and future challenges', *Bulletin of the World Health Organisation*, 78(4): 521–34.

Beyrer C., 1998, 'Burma and Cambodia: human rights, social disruption and the spread of HIV/AIDS', *Health and Human Rights*, 2(4): 85–96.

Bezruchka S., 2000, 'Is globalization dangerous to our health?', *West Journal of Medicine*, 172: 332–4.

Bijlmakers L., M.T. Bassett and D. Sanders, 1998, 'Socio-economic stress, health and child nutritional status in Zimbabwe at a time of economic structural adjustment – a three year longitudinal study', Research Report No. 105 (Uppsala, Sweden: Nordiska Afrikaininstitutet).

Bijlmakers L., M.T. Bassett and D. Sanders, 1999, 'Socio-economic stress, health and child nutritional status at a time of economic structural adjustment – a six year longitudinal study in Zimbabwe', unpublished report.

Birnie P. and A. Boyle, 1992, *International Law and the Environment* (Oxford: Clarendon Press).

Blane R., E. Brunner and R. Wilkinson, eds, 1996, *Health and Social Organisation* (London: Routledge).

Bodansky D., 1995, 'Customary (and not so customary) international environmental law', *Indiana Journal of Global Legal Studies*, 3: 105–19.

Boland L., 1999, 'USG Position on Compulsory Licensing of Patents', 26 March, available on http://www.haiweb.org/campiagn/cl/boland.html.

Bond P., 1999, 'Globalization, pharmaceutical pricing, and South African health policy: managing confrontation with US firms and politicians', *International Journal of Health Services*, 29(4): 765–92.

Booker S., 1999, 'Letter to APIC members', 13 June, Africa Action/Africa Policy Information Center (Electronic Distribution List), Washington DC.

Booth W., 1991, 'Cholera's mysterious journey north', *Washington Post*, 26 August.

Braithwaite J. and P. Drahos, 2000, *Global business regulation* (Cambridge: Cambridge University Press).

Bratton M., 1989, 'The politics of government–NGO relations in Africa', *World Development*, 17(4): 569–87.

Bray R.S., 1996, *Armies of Pestilence: The Effects of Pandemics on History* (Cambridge: Lutterworth Press), pp. 180–3.

Bremner C., J. Sherman and I. Murray, 1997, 'Blair accused of wrecking tobacco ban', *The Times*, 6 November.

Breman A. and Shelton C., 2001, 'Structural adjustment and health: A literature review of the debate, its role players and the presented empirical evidence' WHO Commission on Macroeconomics and Health, Geneva, Working Paper No. WG6: 6.

Broughton M., 1999, Personal communication between Mr Broughton and Dr G.H. Bruntland, WHO Director General, 9 February, cited in Yach and Bettcher (1999).

Broughton M., 2000, 'Doing Business in a Borderless World', speech, International Chamber of Commerce, 33rd World Congress, Budapest, Hungary http://www.bat. com (accessed 11 August 2001).

Broughton M., 2001, 'Doing Business in a Borderless World', Speech to International Chamber of Commerce 33rd World Congress, Budapest, 5 May http://www.bat. com (accessed 11 August 2001).

Brown & Williamson, 1994, 'B&W: US International Brand Strategies', Guildford Depository, Bates No.: 503881246-1270.

Brown & Williamson, 2001, 'Corporate social responsibility', http://www.brownand-williamson.com (accessed 11 August 2001).

Brown P., R.G. Will, R. Bradley, D.M. Asher and L. Detwiler, 2001, 'Bovine Spongiform Encephalopathy and Variant Creutzfeldt-Jakob Disease: Background, evolution, and current concerns', *Emerging Infectious Diseases*, 7(1): 6–16.

Brownlie I., 1998, *Principles of Public International Law*, 5th edn. (Oxford: Clarendon Press).

Bruce-Chwatt L., 1973, 'Global problems of imported disease', *Advances in Parasitology*, 11: 86.

Brugha R. and G. Walt, 2001, 'A global health fund: a leap of faith?', *British Medical Journal*, 323: 152–4.

Brundtland G., 1998, 'WHO Boss sets out Stance on Health and Human Rights', WHO Press Release, 8 December.

Brundtland G., 2001, '24th Session, Codex Alimentarius Commission', WHO speech, Geneva, 2 July, http:/www.who.int/director-gener … dexalimentariuscommission. en.html.

BSE Inquiry Report, 2000, http://www.bseinquiry.gov.uk/index.html.

Buchanan S., 1987, 'Lack of AIDS blood screening in some nations raises concerns', *International Herald Tribune*, 29 September 1987.

Burgess G., 1994, 'Note to Miss H.C. Barton, BAT Industries, Re: TSG – 10–11 May 1994', (plus attached inputs), Guildford Depository, 3 May, Bates No.: 502609233-9252.

Buse K., 1994, 'Spotlight on international organisations: The World Bank', *Health Policy and Planning*, 9: 95–9.

Buse K. and G. Walt, 2000a, 'Global public–private partnerships: Part I – A new development in health?, *Bulletin of the World Health Organization*, 78(4): 549–61.

Buse K. and G. Walt, 2000b, 'Global public-private partnerships: Part II – What are the health issues for global governance?', *Bulletin of the World Health Organization*, 78(5): 699–709.

Butler D. and D. Spurgeon, 1997, 'Canada and France fall out over the risks of asbestos', *Nature*, 358: 379.

Cadwell J., 1993, 'African families and AIDS: Context, reactions and potential interventions', *Health Transition Review*, 3: 1–16.

Cadwell J., 1995, 'Understanding the AIDS epidemic and reacting sensibly to it', *Social Science and Medicine*, 41(3): 301.

Cairncross R., 1997, *The Death of Distance* (Cambridge: Harvard Business School Press).

Caldwell J.C., 1989, 'Mass education as a determinant of mortality decline', in J.C. Caldwell and G. Santow, eds, *Selected Readings in the Cultural, Social and Behavioural Determinants of Health*, Health Transition Series No. 1 (Canberra: Health Transition Centre, Australian National University).

Caldwell J., 2000, 'Rethinking the African AIDS epidemic', *Population and Development Review*, 26(1): 117–35.

Callard C., H. Chitanondh and R. Weissman, 2001, 'Why trade and investment liberalisation may threaten effective tobacco control efforts', *Tobacco Control*, 10: 68–70.

Cardelle A., 1997, 'Health care in the time of reform: emerging policies for private–public sector collaboration in health', *North–South Issues*, 6(1): 1–8.

Challu P., 1991, 'The consequences of pharmaceutical product patenting', *World Competition*, 15: 65–126.

Chaloupka F., T. Hu, K. Warner, R. Jacobs and A. Yurekli, 2000, 'The taxation of tobacco products', in P. Jha and F. Chaloupka, eds, *Tobacco Control in Developing Countries* (Oxford: Oxford University Press), pp. 237–72.

Chase J., 1988, *America Invulnerable: The quest for absolute security from 1812 to Star Wars* (New York: Summit Books).

Chemical and Biological Arms Control Institute/Center for Strategic and International Studies, 2000, *Conflict and Contagion: Health as a Global Security Challenge* (Washington DC: CBACI/CSIS).

Chirimuuta R. and R. Chirimuuta, 1989, *AIDS, Africa and Racism* (London: Free Association).

Chopra M. and D. Sanders, 1997, 'Is growth monitoring worthwhile in South Africa?' *South African Medical Journal*, 87: 875–8.

Chopra M., D. Sanders, D. McCoy and K. Cloete, 1998, 'Implementation of primary health care: package or process?', *South African Medical Journal*, 88(12): 1563–5.

Chugh K., 1992, 'Letter to Honourable Shri ML Fotedar MP, Union Minister for Health, India, Re: Tobacco Policy and the Example of Japan', Guildford Depository, 22 December, Bates No.: 304046975-6981.

Ciresi M., R. Walburn and T. Sutton, 1999, 'Decades of deceit: document discovery in the Minnesota Tobacco Litigation', *William Mitchell Law Review*, 25: 477–566.

Clark A.M., E. Friedman and K. Hochstetler, 1998, 'The sovereign limits of global civil society', *World Politics*, 51: 1–35.

Clark R., 1997, 'Global life systems: biological dimensions of globalisation', *Global Society*, 11(3): 280.

Cliff A. and P. Haggett, 1995, 'Disease implications of global change', in R.J. Johnston, P.J. Taylor and M.J. Watts, eds, *Geographies of Global Change: Remapping the World in the Late 20th Century* (Oxford, England: Blackwell), p. 209.

Climate Change Convention, 1992, 'United Nations Framework Convention on Climate Change', *International Legal Materials*, 31: 849–73.

Codex Alimentarius Commission, 2001a, Codex Alimentarius Commission discusses safety of genetically modified foods, approves toxin limits and guidelines for organic livestock farming, Press Release 6 July 2001, http://www.who.int/inf-pr-2001/en/pr2001-33.html. The report of the 24th session of the Codex commission, www.codexalimentarius.net.

Codex Alimentarius Commission, 2001b, Joint FAO/WHO Food Standards programme. Codex Committee on General Principles. 16th Session. Paris, France 23–27 April 2001. CX/GP 01/3. Agenda item 3a.

Cohen J., 1994, 'AIDS vaccines: Are researchers racing toward success, or crawling?', *Science*, 265: 1373–5.

Cohen M.N., 1989, 'The history of infectious disease', in M.N. Cohen, *Health and the Rise of Civilization* (New Haven: Yale University Press).

Collin J., K. Lee and K. Bissell, 2002, 'The Framework Convention on Tobacco Control: the politics of global health governance', *Third World Quarterly*, 23(2): 265–82.

Collott R., 2001a, 'Brazil and Roche agree deal on AIDS drug price cut', *Financial Times*, 31 August.

Committee of Experts, 2000, 'Tobacco Company Strategies to Undermine Tobacco Control Activities at the World Health Organization', July (Geneva: WHO).

Commonwealth Secretariat, 1989, *Engendering adjustment for the 1990s* (London: Commonwealth Secretariat Publications).

Cookson C., 2001, 'Number of new CJD cases is rising by up to 30% a year', *The Financial Times*, 7 September.

Cookson S., R. Waldman, B. Gushulak *et al.*, 1998, 'Immigrant and refugee health', *Emerging Infectious Diseases*, 4(3): 427–8.

Cooperrider D. and J. Dutton, eds, 1999, *Organizational Dimensions of Global Change* (Thousand Oaks: Sage).

Cornia G.A., 1999, *Liberalization, globalisation and income distribution*, WIDER Working Papers 157 (Helsinki: Wider).

Correa C.M., 2000a, *Intellectual property rights, the WTO and developing countries* (London: Zed Books and Third World Network).

Correa C.M., 2000b, 'Implementing national public health policies in the framework of WTO agreements', *Journal of World Trade*, 34(5): 89–121.

Costello A., F. Watson and D. Woodward, 1994, *Human face or human façade? Adjustment and the health of mothers and children* (London: Centre for International Child Health, University of London).

Coulter A. and C. Ham, eds, 2000, *The Global Challenge of Health Care Rationing* (Buckingham: Open University Press).

Cox R.W., 1987, *Production, Power, and World Order: Social Forces in the Making of History* (New York: Columbia University Press).

Crane S. and J. Carswell, 1992, 'A review and assessment of non-governmental organization-based STD/AIDS education and prevention projects for marginalized groups', *Health Education Research*, 7: 175–94.

Crescenti M., 1999, 'The new tobacco world', *Tobacco Journal International*, March: 51–3.

Crossette B., 1998, 'Kofi Annan's astonishing facts', *New York Times*, 27 September, p. WK16.

Crossette B., 2001, 'For the first time, US is excluded from UN Human Rights Panel', *New York Times*, 4 May 2001.

CSO, 1998, *Poverty in Zimbabwe* (Harare: Central Statistical Office).

Cuddington J., 1993, 'Modelling the macroeconomic effects of AIDS, with an application to Tanzania', *World Bank Economic Review*, pp. 173–89.

Cunningham R., 1996, *'Smoke and Mirrors: The Canadian Tobacco War'* (Ottawa: International Development Research Centre).

Danziger R., 1996, 'HIV/AIDS in the former Soviet Union', unpublished paper, London School of Hygiene and Tropical Medicine, February.

Davis N., 1994, 'Letter to M.F. Broughton, Re: India visit', 5 April 1994, Bates No.: 500030347-35.

Daynard D., C. Bates and N. Francey, 2000, 'Tobacco litigation worldwide', *British Medical Journal*, 320: 111–13.

Dealler S., 1996, *Lethal Legacy, BSE – The Search for the Truth* (London: Bloomsbury).

Dealler S., 2001, 'At long last, signs of a BSE breakthrough', *Guardian*, 5 September: 16.

Decosas J. and J. Finlay, 1993, 'International AIDS aid: The response of development Aid agencies to the HIV/AIDS pandemic', *AIDS*, 7(1): 282–3.

Decosas J., F. Kane, J. Anarfi, K. Sodji and H. Wagner, 1995, 'Migration and AIDS', *The Lancet*, 346: 826.

de Garine I. and N. Pollock, eds, 1995, *Social Aspects of Obesity* (South Australia: Gordon and Breach).

Desai M., 1999, 'From Vienna to Beijing', in P. van Ness, *Debating Human Rights* (London: Rutgers), pp. 184–96.

DHS, 1998, http://www.measuredhs.com

Dicken P., 1998, *Global Shift, Transforming the World Economy* (London: Paul Chapman).

Dollar D., 2001, 'Is globalisation good for your health?', *Bulletin of the World Health Organisation* 79(9): 827–33.

Donahue J. and M. McGuire, 1995, 'The political economy of responsibility in health and illness', *Social Science and Medicine*, 40(1): 47–53.

Donnelley C. and N. Ferguson, 1999, *Statistical Aspects of BSE and vCJD: Model for Epidemics* (London: Chapman & Hall).

Doyal L., 1979, *The Political Economy of Health* (London: Pluto Press).

Doyal L., J. Naidoo and T. Wilton, eds, 1994, *AIDS: Setting a Feminist Agenda* (London: Taylor & Francis).

Drahos P., 1995, 'Global property rights in information: The story of TRIPS at the GATT', *Prometheus*, 13: 6–19.

Drahos P., 1997, 'Thinking strategically about intellectual property rights', *Telecommunications Policy*, 21(3): 201–11.

Drainville A.C., 1998, 'The fetishism of global civil society', in M.P. Smith and L.E. Guarnizo, eds, *Transnationalism from Below* (London: Transaction Publishers), p. 37.

Drewnowski A. and B.M. Popkin, 1997, 'The nutrition transition: new trends in the global diet', *Nutrition Reviews*, 55: 31–43.

Dreyfus R., 1999, 'Big Tobacco Rides East', MOJOwire magazine, January/February http://www.motherjones.com/mother_jones/JF99/dreyfuss.html (accessed 11 August 2001).

Drope J. and S. Chapman, 2001, 'Tobacco industry efforts at discrediting scientific knowledge of environmental tobacco smoke', *Journal of Epidemiology and Community Health*, 55: 588–94.

Duckett M./ICASO, 1999, 'Compulsory Licensing and Parallel Importing', ICASO Background Paper, July.

Duffield M., 1994, 'Complex emergencies and the crisis of developmentalism', *IDS Bulletin*, 25(4): 38.

Dumble L., 1997, 'From mad cows to humans: the next global plague?', *Nexus Magazine*, 5(1), December 1997–January 1998 www.nexusmagazine.com.

Dumble L., 2001, 'The world faces a pandemic of mad cow disease that may rival HIV', *Sydney Morning Herald Section*, 17 January: 12.

Dyer C., 1998, 'Tobacco company set up network of sympathetic scientists', *British Medical Journal*, 316: 1553.

ECJ, 1999, 'Pfizer Animal Health SA/NV v. Council of the European Union', Case T-13/99R, Court of First Instance, 30 June 1999.

The Economist, 1999, 'Mandela's heir', 29 May.

The Economist, 2001, 'The line of least resistance', *The Economist*, 5 May: 71–2.

Eela, 2001, National Veterinary and Food Research Institute, Web-pages: http://www.eela.fi (accessed 11 September 2001).

Einhorn B., 1993, *Cinderella Goes to Market: Citizenship, Gender and Women's Movements in East Central Europe* (London: Verso).

EMRO, 2001, *Voice of Truth*, vol. 2, http://www.emro.who.int/TFI/VoiceOfTruthVol2.pdf (accessed 14 August 2001).

Epstein H., 2001a, 'Time of indifference', *New York Review of Books*, April, 12: 33–8.

Epstein H., 2001b, 'AIDS: The lesson of Uganda', *New York Review of Books*, 5 July: 18–23.

Epstein P., 1995, 'Emerging diseases and ecosystem instability: new threats to public health', *American Journal of Public Health*, 85(2): 168–72.

EU, 1997, Decision 97/87/EC (1997) Official Journal L356.

EU, 1998, Decision 98/84/EC (1998) Official Journal L15.

European Commission, 2000a, Communication on the Precautionary Principle, COM(2000)1 Final of February 2000.

European Commission, 2000b, EC Approach to Government Procurement, April, available from http://www.europa.eu.int/comm/trade/miti/gov_proc/seaproc.html (accessed 11 September 2001).

European Commission, 2001, Treaty of Nice, Amending the Treaty on European Union, The treaties establishing the European Communities and certain related acts, *Official Journal of the European Communities* 10.3.2001.2001/c80/01.

European Court of Justice, 2000, Opinion of Advocate General Fennelly, delivered on 15 June 2000, Case C-376/98, Case C-74/99 European Parliament and Council of the European Union, *The Queen* v. *Secretary of State for health and others*, exp parte Imperial Tobacco Ltd and others.

European Court of Justice, 2001, Case C-405/98, *Konsumentombudsmannen* v. *Gourmet International Products Aktiebolag*, Judgement of the court, 8 March 2001.

Evans R.G., 1997, 'Going for the gold. The re-distributive agenda behind the market-based health care reform', *Journal of Health Politics, Policy and Law*, 22: 427–65.

Evans M., R.C. Sinclair, C. Fusimalohi and V. Liava'a, 2001, 'Globalization, diet and health: an example from Tonga', *Bulletin of the World Health Organization*, 79(9): 856–62.

Evian C., 1994, 'AIDS and the socio-economic determinants of the epidemic in Southern Africa – A cycle of poverty', *Journal of Tropical Pediatrics*, 40: 61–2.

Falk R., 1995, *On Humane Governance: Toward a New Global Politics* (Cambridge, England: Polity Press).

Falk R., 1999, *Predatory Globalization* (Oxford: Blackwell).

Fallows J., 1999, 'The political scientist', *New Yorker*, 7 June.

Farmer P., 1996, 'Social inequalities and emerging infectious diseases', *Emerging Infectious Diseases*, 2(4): 259–66.

FDA, 2000a, 'Labeling Requirements for Systemic Antibacterial Drug Products Intended for Human Use', 65 Federal Register 65611-01, 19 September 2000.

FDA, 2000b, 'A Proposed Framework for Evaluating and Assuring the Human Safety of the Microbial Effects of Antimicrobial New Animal Drugs Intended for Use in Food-Producing Animals', at http://www.fda.gov/cvm/index/vmac/antimi18.html.

Feachem R.G.A., 2001, 'Globalisation is good for your health, mostly', *British Medical Journal*, 323: 504–6.

Fee E. and T.M. Brown, 2001, 'Preemptive biopreparedness: can we learn anything from history?', *American Journal of Public Health*, 91: 721–6.

Fidler D.P., 1998, 'Legal issues associated with antimicrobial drug resistance', *Emerging Infectious Diseases*, 4: 169–77.

Fidler D.P., 1999, *International Law and Infectious Diseases* (Oxford: Clarendon Press).

Fidler D.P., 2001, 'Challenges to humanity's health: the contributions of international environmental law to national and global public health', *Environmental Law Reporter*, 31: 10048–78.

Field M., 1994, 'The health crisis in the Former Soviet Union: A report from the "post war" zone', *Social Science and Medicine*, 41(11): 1469–78.

Final report for the DG Trade of European Commission submitted by CEAS Consultants in Association with Geoff Tansey and Queen Mary Intellectual Property Research Institute, 2000, *Study on the relationship between the Agreement on TRIPS and biodiversity related issues*, September 2000, available from the EU webpages.

Financial Times, 2000, 'Counting the economic costs of AIDS', 17 April.

Finger J.M. and P. Schuler, 1999, *Implementation of Uruguay Round Commitments: The Development Challenge* (Washington, DC: World Bank).

Firn D. and V. Griffith, 2001b, 'Medical journals join forces to issue drug trials warning', *Financial Times*, 10 September.

Fitzroy H., A. Briend and V. Fauveau, 1990, 'Child survival: Should the strategy be redesigned? Experience from Bangladesh', *Health Policy and Planning*, 5: 226–34.

Ford N. and S. Koetsawang, 1991, 'The socio-cultural context of the transmission of HIV in Thailand', *Social Science and Medicine*, 33(4): 405–14.

Fortin A., 1990, 'AIDS, development, and the limitations of the African state', in B. Misztal and D. Moss, eds, *Action on AIDS: National Policies in Comparative Perspective* (New York and London: Greenwood Press).

Foster S. and A. Buve, 1995, 'Benefits of HIV screening of blood transfusions in Zambia', *The Lancet*, 346: 225–7.

Fox N., 1997, 'The madness behind mad cows', in *Spoiled, Why our food is making us sick and what can we do about it* (London: Penguin), pp. 291–331.

Framework Convention Alliance, 2001, 'Briefing Paper for the 2nd Meeting of the Intergovernmental Negotiating Body of the Framework Convention on Tobacco Control: Comments on the Chair's text', March http://www.fctc.org/FCTCfca.shtml (accessed 14 August 2001).

Frank R., 1996, *Virus X: Understanding the Real Threat of the New Pandemic Plagues* (London: HarperCollins).

Frankel G., 1996, 'U.S. aided cigarette firms in conquests across Asia', *Washington Post*, 17 November.

Frenk J., J.L. Bobadilla, J. Sepulveda and M. Lopez Cervantes, 1989, 'Health transition in middle-income countries: new challenges for health care', *Health Policy and Planning*, 4(1): 29–39.

Gadomski A., R. Black and W.H. Mosley, 1990, 'Constraints to the potential impact of child survival in developing countries', *Health Policy and Planning*, 5(3): 235–45.

Gaffney D., A. Pollock, D. Price and J. Shaoul, 1999a, 'NHS capital expenditure and the private finance initiative – expansion or contraction', *British Medical Journal*, 319: 48–51.

Gaffney D., A. Pollock, D. Price and J. Shaoul, 1999b, 'PFI in the NHS – is there an economic case?', *British Medical Journal*, 319: 116–19.

Gallaher, 2001, http://www.gallaher-group.com (accessed 11 August 2001), and advertisements in *New Statesman*, including 25 June and 2 July.

Gannon J.C., 2000, *Global infectious disease threat and its implications for the United States* (Washington DC: National Intelligence Council) http://www.odci.gov/cia/publications/nie/report/nie99-17d.html.

Garrett L., 1994, *The Coming Plague: Newly Emerging Diseases in a World Out of Balance* (New York: Farrar, Straus & Giroux).

Garrett L., 2000a, 'You just signed his death warrant', *Columbia Journalism Review*, November/December, online http://cjr.org/year/00/4/garrett.asp.

Garrett L., 2000b, *Betrayal of Trust: The Collapse of Global Public Health* (New York: Hyperion).

GATT, General Agreement on Tariffs and Trade, as amended 1946, 1966, 1994.

Geiger H.J., 2001, 'Terrorism, biological weapons, and bonanza: assessing the real threat to public health', *American Journal of Public Health*, 91: 708–9.

Gellman B., 2000a, 'Death watch: The belated global response to AIDS in Africa', *Washington Post*, 5 July.

Gellman B., 2000b, 'An unequal calculus of life and death', *The Washington Post*, 27 December.

Giddens A., 1981, *A Contemporary Critique of Historical Materialism* (Berkeley, CA: University of California Press).

Giddens A., 1990, *The Consequences of Modernity* (London: Polity Press).

Giddens A., 1999, 'Runaway world', Reith Lecture Series, BBC, London http://www.bbc.co.uk.

Gill S., 1993, *Gramsci, Historical Materialism and International Relations* (Cambridge: Cambridge University Press).

Gill S., 1995, 'Globalisation, market civilisation, and disciplinary neoliberalism', *Millennium*, 24(3): 399–423.

Gill S. and D. Law, 1988, *The Global political Economy: Perspectives, Problems and Policies* (Hemel Hempstead: Harvester Wheatsheaf).

Gilson L., S. Russell and K. Buse, 1995, 'The political economy of user fees with targeting: developing equitable health financing policy', *Journal of International Development*, 7(3): 369–401.

Glantz S., 2000, 'The truth about big tobacco in its own words', *British Medical Journal*, 321: 313–14.

Glantz S., J. Slade, L. Bero, P. Hanauer and D. Barnes, 1996, *The Cigarette Papers* (Berkeley, CA: University of California Press).

Glass C., 1997, 'The multinationals are coming', *Tobacco Reporter*, January.

Godwin P., 1995, 'Strengthening the role of NGOs in responding to the HIV epidemic in Asia and the Pacific: A UNDP Initiative', *AIDS in Asia Newsletter*, 4: 1, 3.

Gordenker L., R. Coate, C. Jonsson and P. Soderholm, 1995, *International Cooperation in Response to AIDS* (London: Pinter).

Government of British Columbia, Canada, 2001, *GATS and Public Service Systems*, Discussion Paper, Ministry of Employment and Investment, 2 April, available at http://www.ei.gov.bc.ca/trade&export/FTAA-WTO/WTO/governmentalauth.html.

Grady C., 1995, *The Search for an AIDS Vaccine: Ethical Issues in the Development and Testing of a Preventive HIV Vaccine* (Indiana University Press).

Gray N., D. Zaridze, C. Robertson, L. Krivosheeva, N. Sigacheva, P. Boyle and the International Cigarette Variation Group, 2000, 'Variation within global cigarette brands in tar, nicotine, and certain nitrosamines: analytic study', *Tobacco Control*, 9: 351.

Green A. and A. Matthias, 1995, 'NGOs – A policy panacea for the next millenium?', *Journal of International Development*, 7(3): 565–73.

Greenwood J., 1997, *Representing Interests in the European Union* (London: Macmillan).

Greig C., 1993, 'India – Impressions – Letter to N. Davis, BATCo, Staines', Guildford Depository, 27 May, Bates No.: 400464423-4424.

Groennings S., 1995, 'Civil–military alliance to combat HIV and AIDS', *AIDS in Asia*, 4: 8.

GTZ (Deutsche Gessellschaft für Technische Zusammenarbeit), 1995, *Responding to AIDS in the Developing World: The GTZ Contribution* (Eschborn, Germany: GTZ, Health, Population and Nutrition Division).

Guerrant R., 1994, 'Twelve messages from enteric infections for science and society', *American Journal of Tropical Medicine and Hygiene*, 51(1): 27.

Gupta R. *et al.*, 2001, 'Responding to market failures in tuberculosis control', *Sciencexpress*, 19 July 2001 at http://www.sciencexpress.org/19July2001/Page/10.1126/science.1061861.

Haas P.M., 2001, 'International environmental governance: lessons for pollution control since UNCHE', in C. de Jonge Oudraat and P.J. Simmons, eds, *Managing a Globalizing World* (Washington DC: The Carnegie Foundation).

Habermas J., 1984, *Theory of Communicative Action*, volume 1 (Boston: Beacon Press).

Habermas J., 1987, *Theory of Communicative Action*, volume 2 (Boston: Beacon Press).

Hacking I., 1993, 'TSG – Optimising the Management and Effectiveness of the Group's International Motor Sports Sponsorships', BAT documents http://www.library.ucsf.edu/tobacco/batco/html/500/582/otherpages/allpages.html (accessed 25 July 2001).

Hall A.J. and F.T. Cutts, 1993, 'Lessons from measles vaccination in developing countries', *British Medical Journal*, 307: 1294–5.

Hall M., 2001, '£40 m a year to keep "BSE mountain" safe', *The Sunday Telegraph*, 20 May: 17.

Halstead S.B., J.A. Walsh and K. Warren, eds, 1985, *Good health at low cost* (New York: Rockefeller Foundation).

Hammond R., 1998, 'Consolidation in the tobacco industry', *Tobacco Control*, 7: 426–8.

Hanauer P., J. Slade, E. Barnes, L. Bero and S. Glantz, 1995, 'Lawyer control of internal scientific research to protect against product liability lawsuits: the Brown and Williamson documents', *Journal of the American Medical Association*, 274: 234–40.

Handy C., 1995, *Beyond Certainty* (London: Hutchinson).

Harding I., 1989, 'BARCLAY Formula 1 Sponsorship', Guildford Depository, 5 July, Bates No.: 301706447-6454.

Harper T., 2001, 'Ad Watch: Marketing life after advertising bans', *Tobacco Control*, 10: 196–7.

Hastings G. and L. MacFadyen, 2000, 'Keep Smiling, No One's Going to Die: An Analysis of Internal Documents from the Tobacco Industry's Main UK Advertising Agencies' (London: Tobacco Control Resource Centre and the Centre for Tobacco Control Research).

Hawkes S. and K. McAdam, 1993, 'AIDS in the developing world', *Medicine International*, 21(2): 69–72.

Hawthorne P., 1999, 'Let's make an arms deal', *Time Magazine*, 4 October.

Hays J.N., 1998, 'New diseases and transatlantic exchanges', in J.N. Hays, *The Burdens of Disease, Epidemics and Human Response in Western History* (New Brunswick, NJ: Rutgers University Press), pp. 62–77 especially p. 72.

Health Action International, 2000, 'Improving Access to Essential Medicines: Confronting the Crisis' Drug Policy at the 53rd World Health assembly, May 2000 at http://www.haiweb.org/news/WHA53en.html.

Health GAP Coalition, 2000, 'Questioning the UNAIDS/Pharmaceutical Initiative, Seven months and counting...', Position paper presented at UNAIDS meeting in Rio de Janeiro, 13 December, available at http://www.globaltreatmentaccess.org/content/camp/.

Health Horizons, 2000, 'Ensuring access in developing countries', No. 39, Summer: 12.

Held D., A. McGrew, D. Goldblatt and J. Perraton, 1999, *Global transformations. Politics, Economics and Culture* (Cambridge: Polity).

Hemminki E., D. Hailey and M. Koivusalo, 1999, 'Battles in court', *Science*, 285: 203–4.

Henretig F., 2001, 'Biological and chemical terrorism defense: a view from the "front lines" of public health', *American Journal of Public Health*, 91: 718–20.

Henry A., 2001, 'Jordan fire Frentzen and turn to Zonta for German grand prix', *Guardian*, 26 July.

Herter U., 2000, 'Industry Perspective: A perspective on the global tobacco market', http://www.bat.com (accessed 5 August 2001).

Heymann D.L. and G.R. Rodier, 1998, 'Global surveillance of communicable diseases', *Emerging Infectious Diseases*, 4(3): 1–5.

Hobsbawm E., 1975, *The Age of Capital: 1848–1875* (London: Weidenfield & Nicolson), p. 310.

Holland W. and S. Stewart, 1997, *Public Health, The Vision and the Challenge* (London: The Nuffield Trust).

Hong E., 2000, 'Globalisation and the impact on health: A third world view', Third World Network, Penang, Malaysia, August, available at http://www.twnside.org.sg/health.html.

Hontiveros-Braquel R., 1996, 'Gender and Security', paper delivered at 10th Asia-Pacific Roundtable, June, Kuala Lumpur.

Hoogvelt A., 1997, *Globalisation and the Postcolonial World: The New Political Economy of Development* (London: Macmillan).

Hurrell A., 1999, 'Security and Inequality', in A. Hurrell and N. Woods, eds, *Inequality, Globalization and World Politics* (Oxford: Oxford University Press).

Hutton W., 1995, *The State We're In* (London: Jonathan Cape).

IAVI, undated, 'Accelerating an AIDS vaccine for developing countries: Recommendations for the World Bank', *Economics of AIDS Vaccines* (New York: International AIDS Vaccine Initiative), http://www.iavi.org/globalmobil_z_economics.html.

IAVI Report, 2000, 'The view from SmithKline: An Interview with Jean Stephenne', *IAVI Report*, April–June www.iavi.org/reports/12/report_june2000_07.html.

Idelovitch E. and K. Ringskog, 1995, *Private Sector Participation in Water Supply and Sanitation in Latin America* (Washington DC: World Bank, Directions in Development Series).

Illich I., 1976, *Limits to Medicine* (London: Penguin).

Imperial Tobacco, 1987, 'TAC Research Committee, Draft Document for discussion', December, Bates No.: 400113687-3693, cited in Drope and Chapman (2001).

INFOTAB, 1987, 'World Action! A Guide for Dealing with Anti-Tobacco Pressure Groups', Guildford depository 23/1/01, File No. CA0080, Box No.: 601005635, Bates No.: 601005624-5635.

Institute of Medicine, 1992, *Emerging Infections: Microbial Threats to Health in the United States* (Washington DC: National Academy Press).

Institute of Medicine, 1997, *America's Vital Interest in Global Health* (Washington DC: National Academy Press).

International Consortium of Investigative Journalists, 2001, 'Tobacco Companies Linked to Criminal Organizations in Cigarette Smuggling', 3 March http://www.public-i.org/story_01_030301.html (accessed 9 August 2001).

ITC, 1994, 'Business Plan – Project Barracuda', Guildford Depository, January, Bates No.: 500350130-0175.

Jacobson P. and K. Warner, 1999, 'Litigation and public health policy-making: the case of tobacco control', *Journal of Health Politics, Policy and Law*, 24(4): 769–804.

Jacobzone S., 2000, *Pharmaceutical policies in OECD countries: reconciling social and industrial goals*, OECD Labour Market and Social Policy – Occasional Papers No. 40. DEELSA/ELSA/WD(2000)1.

James J., 1999a, 'New Frontier of AIDS Activism: International Trade Rules and Global Access to Medicines – Interview with Eric Sawyer, HIV/AIDS Human Rights Project', *Aids Treatment News Archive*, 16 April.

James J., 1999b, 'Compulsory Licensing for Bridging the Gap – Treatment Access in Developing Countries: interview with James Love, Consumer Project on Technology', *AIDS Treatment News Archive*, 5 March http://www.aids.org/Immunet/atn.nsf/page/a-314-01.

James J., 2000, 'Africa Treatment Access in the News', *AIDS Treatment News*, 343, 19 May.

Jameson F. and M. Miyoshi, eds, 1998, *The Cultures of Globalization* (Durham: Duke University Press).

Japan Tobacco International, 2001, 'What we stand for', http://www.jti.com/e/what_we_stand_for/addiction/what_addiction_e.html (accessed 11 August 2001).

Jha P. and F. Chaloupka, 1999, *Curbing the Epidemic: Governments and the Economics of Tobacco Control* (Washington DC: World Bank).

Jones R.J.B., 1995, *Globalisation and Interdependence in the International Political Economy: Rhetoric and Reality* (London: Pinter).

Joossens L., 1998, 'Tobacco smuggling: an optimal policy approach', in I. Abedian, R. van der Merwe, N. Wilkins and P. Jha, eds, *The Economics of Tobacco Control: Towards an optimal policy mix* (Cape Town, South Africa: Applied Fiscal Research Centre, University of Cape Town).

Joossens L. and M. Raw, 1998, 'Cigarette smuggling in Europe: who really benefits?', *Tobacco Control*, 7(1): 66–71.

Joossens L., F. Chaloupka, D. Merriman and A. Yurekli, 2000, 'Issues in the smuggling of tobacco products', in P. Jha and F. Chaloupka, eds, *Tobacco Control in Developing Countries* (Oxford: Oxford University Press), pp. 393–406.

Jutzi S., 2001, 'Market liberalisation could spread mad cow risk: UN official', http://www.purefood.org/madcow/market61101.cfm, 11 June.

Kaferstein K.F., Y. Motarjemi and D. Bettcher, 1997, 'Foodborne disease control: a transnational challenge', *Emerging Infectious Diseases*, 3(4): 1–10.

Kaiser Daily HIV/AIDS Report, 2000a, 'Africa Trade Bill: Clinton Issues Executive Order Relaxing Intellectual Property Rights', 11 May 2000 at http://report.kff.org/archive/aid/2000/05/kh000511.1.html.

Kaiser Daily HIV/AIDS Report, 2000b, 'Africa: Southern countries lack infrastructure to accept offer of price cuts for AIDS drugs', 21 June http://report.kff.org/hivaids.

Kaiser Daily HIV/AIDS Report, 2001, 23 January http://report.kff.org/hivaids.

Kaldor M., 1999, *New and Old Wars* (Cambridge: Polity).

Kane S., 1993, 'Prostitution and the military: Planning AIDS intervention in Belize', *Social Science and Medicine*, 36(7): 965–79.

Kane F., M. Alary and I. N'Doye, 1993, 'Temporary expatriation is related to HIV-1 infection in rural Senegal', *AIDS*, 7: 1261–5.

Karlen A., 1995, *Man and Microbes, Disease and Plagues in History and Modern Times* (New York: Simon & Schuster).

Kassalow J.S., 2001, *Why Health Is Important to U.S. Foreign Policy* (New York: Council on Foreign Relations and Milbank Memorial Fund).

Kaufman N. and M. Nichter, 2001, 'The marketing of tobacco to women: global perspectives', in J. Samet and S. Yoon, eds, *Women and the Tobacco Epidemic: Challenges for the 21st Century*, WHO/NMH/TFI/01.1 (WHO: Geneva), pp. 69–98.

Kaul I., I. Grunberg and M.A. Stern, 1999, 'Defining global public goods', in I. Kaul, I. Grunberg and M.A. Stern, eds, *Global Public Goods, International Cooperation in the 21st Century* (Oxford: Oxford University Press). pp. 2–19.

Kennedy M., 2001, 'A terrible killer', *Times Colonist*, 3 June: D3.

Keohane R.O., 2001, 'Governance in a partially globalized world', *American Political Science Review*, 95: 1–13.

Khor M., 1996, 'Health: BSE – False alarm or calamity?', *Third World Network*, 10 April.

Kickbusch I.S., 2000, 'The development of international health policies – accountability intact?', *Social Science and Medicine*, 51: 979–89.

Kickbusch I.S., 2001, 'Health literacy: addressing the health and education divide', *Health Promotion International*, 16: 289–97.

Kickbusch I.S. and K. Buse, 2000, 'Global influences and global responses: international health at the turn of the twenty-first century', in M.H. Merson, R.E. Black and A.J. Mills, eds, *International Public Health: Diseases, Programs, Systems, and Policies* (Maryland: Aspen Publishers).

Kidder T., 2000, 'The good doctor', *New Yorker*, 10 July: 40–57.

Kim J., J. Millen, A. Irwin and J. Gershman, eds, 2000, *Dying for Growth: Global Inequality and the Health of the Poor* (Maine: Common Courage Press).

Kinnon C., 1998, 'Globalization, world trade: bringing health into the picture', *World Health Forum*, 19: 397–406.

Kiple K., ed., 1993, *The Cambridge World History of Human Disease* (Cambridge: Cambridge University Press), pp. 642–9.

Klaukka T. and S. Rajaniemi, 2001, 'Mitkä lääkeryhmät kasvattivat eniten korvausmenoja vuonna 2000?', *TABU*, 3: 20–2.

Klein N., 2000, *No Logo* (London: Flamingo).

Kluger A., 1996, *Ashes to Ashes: America's hundred-year war, the public health, and the unabashed triumph of Philip Morris* (New York: Alfred A. Knopf).

Koivusalo M., 1999, *The World Trade Organisation and trade-creep in health and social policies*, GASPP Occasional Papers No. 4/1999 (Helsinki: STAKES), available at http://www.stakes.fi/gaspp/.

Koivusalo M. and M. Rowson, 2000, 'The WTO: implications for health policy', *Medicine, Conflict and Survival*, 16: 175–91.

Krajewski M., 2001, Public services and the scope of the General Agreement of Trade in Services, a research paper written for Center for International Environmental Law (CIEL), May 2001.

Kuhn T., 1996, *The Structure of Scientific Revolutions*, 3rd edn (Chicago: University of Chicago Press).

Kuritzkes D.R., 2000, Update on HIV Drug Resistance Information Presented at 40th ICAAC (17–20 Sept. 2000), at http://www.hivandhepatitis.com/hiv/v09270001.html.

Kyoto Protocol, 1997, 'Kyoto Protocol to the United Nations Framework Convention on Climate Change', UN Doc. FCCC/CP/1997/L.7/Add.1.

Labonte R., 1998, 'Healthy public policy and the WTO: a proposal for an international health presence in future trade/investment talks', *Health Promotion International*, 13(3): 245–56.

Laing R., 1999, 'Global Issues of access to Pharmaceuticals and Effects of Patents' presentation to the AIDS and Essential Medicines and Compulsory Licensing Meeting, Geneva, 26 March http://www.haiweb.org/campaign/cl/laing.html.

Lal S., 1995, 'The HIV/AIDS situation in India', *AIDS in Asia*, 4: 6.

Lancet, The, 1994, AIDS: The third wave', *The Lancet*, 343: 186–8.

Lang T., 1999, 'Diet, health and globalisation: five key questions', International and Public Health Group Symposium on 'Feeding the world in the future', *Proceedings of the Nutrition Society*, 58: 335–43.

Lee K., 1995, 'A neo-Gramscian approach to international organization: an expanded analysis of current reform to UN development activities', in J. MacMillan and A. Linklater, eds, *Boundaries in Question: New Directions in International Relations* (London: Pinter), pp. 142–62.

Lee K., 1998a, 'Globalisation and Health Policy, A Review of the Literature and Proposed Research and Policy Agenda', LSHTM Discussion Paper, London.

Lee K., 1998b, 'Shaping the future of global health cooperation: where can we go from here?', *The Lancet*, 351: 899–902.

Lee K., 2000, 'Globalisation and health policy: a review of the literature and proposed research and policy agenda', in A. Bambas, J.A. Casas, H. Drayton and A. Valdes, eds, *Health & Human Development in the New Global Economy*, (Washington DC: Pan American Health Organization), pp. 15–42.

Lee K., 2001, 'A dialogue of the deaf? The health impacts of globalisation', *Journal of Epidemiology and Community Health*, 55(9): 619.

Lee K., 2003, *Globalization and Health, An Introduction* (London: Palgrave).

Lee K. and R. Dodgson, 1998, 'Globalisation and Cholera: Implications for Global Governance', Paper presented to the British International Studies Association Annual Conference, University of Sussex, Falmer, December.

Lee K. and A. Zwi, 1996, 'A global political economy approach to AIDS: Ideology, interests and implications', *New Political Economy*, 1(3): 355–73.

Legge D.M., 1993, 'Investing in the shaping of world health policy', Paper presented at AIDAB, NCEPH and PHA Workshop to discuss 'Investing in Health', Canberra, Australia, 31 August.

Leppo K., 1997, 'Introduction', in M. Koivisalo and E. Ollila, eds, *Making a Healthy World* (Finland: Zed Books).

Lightwood J., D. Collins, H. Lapsley and T. Novotny, 2000, 'Estimating the costs of tobacco use', in P. Jha and F. Chaloupka, eds, *Tobacco Control in Developing Countries* (Oxford: Oxford University Press), pp. 63–103.

Lister C., 1993, 'Memorandum to Mr Pantos Re: Cairo IAQ Conference', 26 October, Bates No.: 202580647-0648 http://www.pmdocs.com, cited: EMRO (2001).

Lob-Levyt J., 1990, 'Compassion, economics, politics ... What are the motives behind health-sector aid?', *Health Policy and Planning*, 5(1): 82–7.

Love J., 1999a, 'Five Common Mistakes by reporters covering US/South Africa disputes over compulsory licensing and parallel imports', 23 September http://www.cptech.org/ip/health/sa/mistakes.html.

Love J., 1999b, 'What is the United States' Role in Combating the Global HIV/AIDS Epidemic?', Statement of James Love before the Subcommittee on Criminal Justice, Human resources and Drug Policy, Committee on Government Reform, US, 22 July http://legalminds.findlaw.com/list/info-policy-notes/msg00093.html.

Lovett N., 1992, 'Fax to S. Cook, Subject – Draft Health Policy Warning', Guildford Depository, 2 April, Bates No.: 301643951-3954.

Lurie P., P. Hintzen and R. Lowe, 1995, 'Socioeconomic obstacles to HIV prevention and treatment in developing countries: the roles of the International Monetary Fund and the World Bank', *AIDS*, 9: 539–46.

Maguire K., 2000, 'A tobacco giant and its global reach', *Guardian*, 31 January http://www.guardian.co.uk/bat/article/0,2763,191299,00.html (accessed 8 August 2001).

Makinda S., 2000, 'Recasting global governance', in R. Thurkur and E. Newman, eds, *New Millennium, new perspectives: The United Nations, security, and governance* (Tokyo, New York, Paris: United Nations University Press).

Malloch Brown M., 2000, Statement at UN Security Council Meeting, 10 January.

Manivannan G. and S.P. Sawan, 2000, 'Perspectives on infectious disease challenges and antimicrobial answers', in S.P. Sawan and G. Manivannan, eds, *Antimicrobial/Anti-infective materials: principles, applications and devices* (Lancaster, PA: Technomic Publishing Co., Inc.), pp. 1–22.

Mankahlana P., Head of Communications, President Mbeki's office, 2000, Press release, 24 March.

Mann J., 1988, 'The Global Picture of AIDS', address to the Fourth International Conference on AIDS, Stockholm, 12 June, pp. 4–5.

Mann J. and K. Kay, 1991, 'Confronting the Pandemic: the WHO's GPA 1986–9', *AIDS*, 5(Supplement 2): S221–9.

Mann J. and D. Tarantola, 1995, 'Preventative medicine: A broader approach to the AIDS crisis', *Harvard International Review*, 17(4): 46–9, 87.

Mann J., D. Tarantola and T. Netter, eds, 1992, *AIDS in the World: A Global Report* (Boston, MA: Harvard University Press).

Maragos J.E., 1994, 'Description of reefs and corals for the 1988 Protected Area Survey of the Northern Marshall Islands', *Atoll Research Bulletin*, 419: 1–84.

Marais H., 2000, 'To the edge', *AIDS Review 2000*, University of Pretoria.

Marceau G. and Pedersen P.N., 1999, 'Is the WTO open and transparent? A discussion of the relationship of the WTO with non-governmental organisations and civil society's claims for more transparency and public participation', *Journal of World Trade*, 33(1): 5–49.

Massing M., 2000, 'The Narco-State', *New York Review of Books*, 15 June, 29: 26.

Maxwell R., 1997, 'An unplayable hand?' *BSE, CJD and British Government* (London: King's Fund).

McCarthy M., 1995, 'New US Head of HIV/AIDS Policy appointed', *The Lancet*, 346: 62.

McCarthy M., 1998, 'Public rejects genetically modified food', *The Independent*, 18 November: 5.

McCarthy S., R. McPhearson, A. Guarino and J. Gaines, 1992, 'Toxigenic vibrio cholerae 01 and cargo ships entering Gulf of Mexico', *The Lancet*, 339: 624–5.

McMichael A.J., B. Bolin, R, Costanza, G. Daily, C. Folke, K. Lindahl-Kiessling, E. Lindgren and B. Niklasson, 1999, 'Globalization and the sustainability of human health: an ecological perspective', *BioScience*, 49(3): 205–10.

McMurray C. and R. Smith, 2001, *Diseases of Globalization* (London: Earthscan).

Meikle J., 2001, 'Food agency on trail of low grade meat', *Guardian*, 10 August: 7.

Meikle J. and A. Bellos, 2001, 'CJD link to blood Britain sold abroad', *Guardian*, 5 February.

Merson M., 1993, 'Slowing the spread of HIV: Agenda for the 1990s', *Science*, 260: 1266–8.

Messner D., 1999, 'Globalisierung, global governance und entwicklungspolitik', in *Internationale Politik*, 1, S.5–18.

Michaud C. and C. Murray, 1994, 'External assistance to the health sector in developing countries: a detailed analysis 1972–90', *Bulletin of the World Health Organization*, 72: 639–51.

Millen, J., A. Irwin, J. Kim and J. Gershman, 2000, *Dying for Growth: Global Inequality and the Health of the Poor* (Marine, MS: Common Courge Press).

Mills S., ed., 1988, *Alternatives in Healing: An Open-minded Approach to Finding the Best Treatment for Your Health Problems* (London: Macmillan).

Misra S., 1988, 'Letter to AWH Suszynski, BATCo Ltd', 18 April, Bates No.: 301531792.

Mohammadi A., ed., 1997, *International Communication and Globalization* (London: Sage).

MOHE, 1996, *Annual Report*, Marshall Islands Government, Ministry of Health and Environment.

MOHE, 1997, *Annual Report*, Marshall Islands Government, Ministry of Health and Environment.

Montreal Protocol, 1987, 'Montreal Protocol on substances that deplete the ozone layer', *International Legal Materials*, 26: 1550–61.

Moon G. and R. Gillespie, eds, 1995, *Society and Health: An Introduction to Social Science for Health Professionals* (London: Routledge).

Moses F., F. Manji and J. Bradley, 1992, 'Impact of user fees on attendance at a referral center for sexually transmitted diseases in Kenya', *The Lancet*, 340: 463–6.

Mosley W. Henry, 1985, Part III Discussion 'Remarks' in S.B. Halstead *et al.*, op. cit., pp. 241–4.

Motarjemi Y. and F. Kaferstein, 1997, 'Global estimation of foodborne diseases', *World Health Statistics Quarterly*, 50(1/2): 5–11.

Mulligan P., 1993, 'Nigeria, Status, February 1993', Guildford Depository, 17 February, Bates No.: 500253754-757.

Murphy C., 2000, 'Political Consequences of the New Inequality', Presidential address to the International Studies Association Annual Conference, Washington DC, February.

Murphy C. and R. Tooze, 1991, *The New International Political Economy* (Boulder, CO: Lynne Rienner).

Murray I., 1998, 'CJD risk threatens ban on British blood', *The Times*, 17 July: 1.

Musasa Project, 1998, 'An Investigation into the Relationship Between Domestic Violence and Women's Vulnerability to Sexually Transmitted Infections and HIV/AIDS', Harare, Zimbabwe.

Narkevich M. *et al.*, 1993, 'The seventh pandemic of cholera in the USSR, 1961–89', *Bulletin of the World Health Organization*, 71(2): 189–96.

Navarro V., 1981, *Imperialism, Health and Medicine* (Amityville, NY: Baywood).

Navarro V., 1998, 'A historical review (1965–1997) of studies on class, health and quality of life: a personal account', *International Journal of Health Services*, 28(3): 389–406.

New Internationalist, 1998, 'Migration: the facts', *New Internationalist*, 305: 18.

Nicholson D., 2001, 'The future of food: safety first for agriculture; the foot-and-mouth and BSE crises have forced a reappraisal of a way Britain produces food', *The Independent*, 7 July.

Nogues J., 1990, 'Patents and pharmaceutical drugs: understanding the pressures on developing countries', *Journal of World Trade Law*, 24: 81–104.

OECD, 1997, The OECD Report on Regulatory Reform, Synthesis (Paris: OECD).

Office of the Spokesperson for the Secretary-General, 'Contributions Pledged to the Global AIDS and Health Fund', accessed through: http://www.un.org/News/ossg/aids.html (18 September 2001).

Oh C., 2000, 'TRIPS and Pharmaceuticals: A case of corporate profits over public health', at http:twnside.org.sg/title/twr120a.html.

Okeke I.N. and R. Edelman, 2001, 'Dissemination of antibiotic-resistant bacteria across geographic borders', *Clinical Infectious Diseases*, 33: 364–9.

Olson E., 2000, 'Worldwide meat trade might have spread disease', *International Herald Tribune*, 23 December.

Ong E. and S. Glantz, 2000, 'Tobacco industry efforts subverting International Agency for Research on Cancer's second-hand smoke study', *The Lancet*, 355(9211): 1253–9.

Orbinski J., 2001, 'Health, equity and trade: A failure in global governance', in G.P. Sampson, ed., *The Role of the World Trade Organization in Global Governance* (Tokyo, New York, Paris: United Nations University Press).

Otsuki T., J.S. Wilson and M. Sewadeh, 2001, 'A Race to the Top? A Case Study of Food Safety Standards and African Exports', Working Paper No. 2563, Development Research Group (Washington, DC: World Bank).

Ourusoff A., 1992, 'What's in a name? What the world's top brands are worth', *Financial World*, 1 September: 32–49.

Oyaku R., 2001, 'Wars and HIV/AIDS spread in sub-Saharan Africa', posting to 'Break the Silence', 19 April break-the-silence@hdnet.org.

Panos Institute, 1988, *AIDS and the Third World* (London: Panos Institute, with the Norwegian Red Cross).

Panos Institute, 2000, *Beyond our Means: the Cost of Treating HIV/AIDS in the Developing World* (London: Panos Institute).

Paquet G., 2001, 'The new governance, subsidiarity and the strategic state, in *Governance in the 21st Century*' (Paris: OECD), pp. 183–214.

Patten C., 2000, 'Governance', Reith Lecture Series, BBC, London http://news.bbc.co.uk/hi/english/static/events/reith_2000/lecture1.stm.

Pearce N., 1996, 'Traditional epidemiology, modern epidemiology, and public health', *American Journal of Public Health*, 86: 678–83.

Perez-Casas C., P. Chirac, D. Berman and N. Ford, 2000, 'Access to Fluconazole in less-developed countries', *The Lancet*, 356(9247): 1584.

Phillips P.W.B. and W.A. Kerr, 2000, 'Alternative Paradigms, The WTO versus the Biosafety Protocol for Trade in Genetically Modified Organisms', *Journal of World Trade*, 34(4): 64–75.

Piot P., 2000, statement to UN Security Council meeting on HIV/AIDS in Africa, 10 January.

Pitt D., 1992, 'Power in the UN superbureaucracy: A new Byzantium?', in D. Pitt and T. Weiss, *The Nature of United Nations Bureaucracies* (London: Croom Helm), chap. 2.

Plotkin B.J. and A.M. Kimball, 1997, 'Designing an international policy and legal framework for the control of emerging infectious diseases: first steps', *Emerging Infectious Diseases*, 3(1): 1–9.

Pollock A., M. Dunnigan, D. Gaffney, D. Price and J. Shaoul, 1999, 'Planning for the new NHS: downsizing for the 21st century', *British Medical Journal*, 319: 1–6.

Pollock A. and D. Price, 2000, 'Rewriting the regulations: How the World Trade Organisation could accelerate privatisation in health-care systems', *The Lancet*, 356: 1995–2000.

Popkin B.M., 1994, 'The nutrition transition in low-income countries: an emerging crisis', *Nutrition Reviews*, 52: 285–98.

Porter M., 1995, 'From Yokohama to Vancouver – Towards the 11th International AIDS Conference', *AIDS in Asia*, 4: 5.

Porter D., 1997, 'A plague on the borders', in L. Manderson and M. Jolley, *Sites of Desire/Economies of Pleasure* (Chicago: University of Chicago Press), pp. 213–14.

Potts M. and W. Carswell, 1993, 'AIDS: Losing the battle and the war?', *The Lancet*, 341: 1442–3.

Powell D. and W. Leiss, 1997, *Mad Cows and Mother's Milk* (Montreal: McGill-Queens University Press).

Preston R., 1994, *The Hot Zone* (New York: Corgi).

Price D., A. Pollock and J. Shaoul, 1999, 'How the World Trade Organisation is shaping domestic policies in health care', *The Lancet*, 354: 1889–92.

Prior P. and R. Sykes, 2001, 'Globalisation and European Welfare States: Evaluating the Theories and Evidence', in R. Sykes, B. Palier and P. M. Prior, eds, *Globalisation and European Welfare States: Challenges and Change* (Houndmills and New York: Palgrave).

Prusiner S., 1998, 'Prions', *Proceeding of the National Academy of Sciences*, online edition, 95(23): 13363–83.

Public Citizen, 2001, *Rx R&D Myths: The Case Against the Drug Industry's R&D "Scare Card"*, 23 July http://www.citizen.org/congress/drugs/R&Dscarecard.html.

Public Citizens Global Trade Watch, 1997, *NAFTA's Broken Promises: Fast Track to Unsafe Food* (Washington DC).

Public Health Improvement Act, 2000, Public Law 106–505, 13 November 2000, §319E – Combating Antimicrobial Resistance; 42 United States Code Annotated §247d-5.

Raghavan C., 1996, 'What is globalisation?', *Third World Resurgence*, 74: 11–14.

Ranson K., P. Jha, F. Chaloupka and S. Nguyen, 2000, 'The effectiveness and cost-effectiveness of price increases and other tobacco control policies', in P. Jha and F. Chaloupka, eds, *Tobacco Control in Developing Countries* (Oxford: Oxford University Press), pp: 427–47.

Ratzan S., ed., 1998, *The Mad Cow Crisis, Health and the Public Good* (London: UCL Press).

Reinicke W.H., 1998, *Global Public Policy: Governing with government?* (Washington, DC: Brookings Institution Press).

Rembiszewski J., 1994, 'Letter to J. Winebrenner, Brown & Williamson, Subject – SE 555 – Japan', Guildford Depository, Bates No.: 500045763-5764.

Reuters Medical News, 2001, Antibiotic Resistance Problem Addressed in Proposed Legislation, 9 May at http://www.id.medscape.com/reuters/prof/2001/05/05.10/20010509legi003.html.

Reynolds C., 1999, 'Tobacco advertising in Indonesia: "the defining characteristics for success"', *Tobacco Control*, 8: 85–8.

Rhodes R., 1997, *Deadly Feasts, The 'Prion' Controversy and the Public's Health* (New York: Simon & Schuster).

Rice T., 1997, 'Can markets give us the health system we want', *Journal of Health Politics, Policy and Law*, 22: 383–426.

Richards T., 2001, 'Editorial: The new global health fund', *British Medical Journal*, 322: 1321–2.

Ricketts M., 2000, 'UN fears BSE may have spread worldwide: 500,000 tons of meal exported', AP Worldstream http://www.mad-cow.org/00/dec00_31_news.html.

Ridley R. and H. Baker, 1998, *Fatal Protein: The Story of CJD, BSE and Other Prion Diseases* (Oxford: Oxford University Press).

Rifkin S. and G. Walt, 1986, 'Why health improves: defining the issues concerning "comprehensive primary health care" and "selective primary health care"', *Social Science and Medicine*, 23: 559–66.

Robertson R., 1992, *Globalization, Social Theory and Global Culture* (London: Sage).

Robinson S., 2001, 'Battle ahead', *Time Magazine* (Australian edition) 16 July: 40–1.

Rockefeller Foundation, 1994, *Accelerating the Development of Preventative HIV Vaccines for the World* (New York: Rockefeller Foundation).

Rogers D., 2001a, 'BAT lures young smokers with "devious" online scheme', *Guardian*, 24 January http://media.guardian.co.uk/marketingandpr/story/0,7494,427372,00. html (accessed 10 August 2001).

Rogers D., 2001b, 'Survey – Creative business: tobacco advertising', *Financial Times*, 24 July.

Rosenau J., 1995, 'Governance in the twenty-first century', *Global Governance*, 1(1): 13–43.

Rosenberg T., 2001, 'The Brazilian solution', *New York Times Magazine*, 28 January: 26–31; 52–5.

Rubenstein D., ed., 1992, *Pacific History: Papers from the 8th Pacific History Association Conference* (Mangilao, Guam: University of Guam and the Micronesian Area Research Center).

Rubenstein K. and D. Adler, 2000, 'International citizenship: The future of nationality in a globalized world', *Indiana Journal of Global Legal Studies*, 7(2): 519–48.

Ruggie M., 1996, *Realignments in the Welfare State: Health Policy in the United States, Britain, and Canada* (New York: Columbia University Press).

Ruggie J.G., 1998, *Constructing The World Polity: Essays in International Relations* (London, Canada, New York: Routledge).

Rugman A. and M. Gestrin, 1997, 'New rules for multilateral investment', *International Executive*, 39(1): 21–33.

Ruohonen M., 1999, 'Sosiaalipalvelussa kilpailu ei toimi', *Helsingin Sanomat*, 18 December.

Rupp J. and D. Billings, 1990, 'Asia ETS Consultant Status Report', 14 February http://www.pmdocs.com, Bates No.: 2500048976-8998.

Russett B. and J. Oneal, 2001, *Triangulating Peace: Democracy, Interdependence and International Organizations* (New York: W.W. Norton Inc.).

Ryan F., 1996, *Virus X, Understanding the Real Threat of the New Pandemic Plagues* (New York: HarperCollins).

Sachs J., 2001, 'The best possible investment in Africa', *New York Times*, op. ed., 10 February.

Saloojee Y. and E. Dagli, 2000, 'Tobacco industry tactics for resisting public policy on health', *Bulletin of the World Health Organization*, 78(7): 902–10.

Sanchez J. and D. Taylor, 1997, 'Cholera', *The Lancet* 349: 1825–30.

Sanders D., 1985, *The Struggle for Health: Medicine and the Politics of Under-development* (London: Macmillan).

Sanders D., 1998, 'PHC 21 – Everybody's business', Main background paper for the meeting: PHC 21 – Everybody's business, An international meeting to celebrate 20 years after Alma-Ata, Almaty, Kazakhstan, 27–28 November 1998, WHO Report WHO/EIP/OSD/00.7 (Geneva: World Health Organization).

Sanders D., 1999, 'Success factors in community-based nutrition programmes', *Food and Nutrition Bulletin*, 20(3): 307–14.

Sanders D. and R. Davies, 1988, 'The economy, the health sector and child health in Zimbabwe since independence', *Social Science and Medicine*, 27(7): 723–31.

Sanders D. and A. Sambo, 1991, 'AIDS in Africa: The implications of economic recession and structural adjustment', *Health Policy and Planning*, 6(2): 157–65.

Sanger M., 2001, *Reckless Abandon: Canada, the GATS and the Future of Health Care* (Ottawa: Canadian Centre for Policy Alternatives).

Sapru J., 1980, 'Smoking and health in the Indian environment', Guildford Depository, Bates No.: 102403605-3622.

Sargent J., J. Tickle, M. Beach, M. Dalton, M. Ahrens and T. Heatherton, 2001, 'Brand appearances in contemporary cinema films and contribution to global marketing of cigarettes', *The Lancet*, 357: 29–32.

Sauve P. and R.M. Stern, eds, 2000, *GATS 2000. New directions in services trade liberalisation* (Washington, DC: Brookings Institution Press).

Schachter O., 1991, 'The emergence of international environmental law', *Journal of International Affairs*, 44: 457–93.

Schneider H. and J. Stein, 2001, 'Implementing AIDS policy in post-apartheid South Africa', *Social Science and Medicine*, 52: 723–31.

Schoeffel P., 1992, 'Food, health and development in the Pacific Islands: policy implications for Micronesia', *ISLA: A Journal of Micronesian Studies*, I(2): 223–50.

Schofield R. and J. Shaoul, 1998, 'E. Coli 0157: Public health vs private wealth', *Eurohealth*, 4(4): 35–7.

Scholte J.A., 1997, 'The globalization of world politics', in J. Baylis and S. Smith, eds, *The Globalization of World Politics: An Introduction to International Relations* (Oxford: Oxford University Press), pp. 13–30.

Scholte J.A., 2000, *Globalization, a critical introduction* (London: Macmillan).

Scholte J.A., R. O'Brien and M. Williams, 1999, 'The WTO and Civil Society', *Journal of World Trade*, 33(1): 107–23.

Schoofs M., 2000a, 'Glaxo attempts to block access to generic Aids drugs in Ghana', *Wall Street Journal*, 1 December.

Schoofs M., 2000b, 'Drug majors battle for Ghana's AIDS market', *Sanchita Sharma Indian Express*, New Delhi, 17 November.

Schovtz M., 1999, 'Views and perspectives on Compulsory Licensing of Essential Medicines', 26 March, at htttp://www.haiweb.org/campaign/cl/scholtz.html.

Schwartlander B. *et al.*, 2001, 'AIDS: Resource needs for HIV/AIDS, *Science*, 292(5526): 2434–6.

Science, 1994, 'Bumps on the vaccine road', *Science*, 265: 1371–3.

Seidel G., 1993, 'The competing discourses of HIV/AIDS in Sub-Saharan Africa: Discourses of rights and empowerment vs discourses of control and exclusion', *Social Science and Medicine*, 36: 175–94.

Sen G. and A. Gurumuthy, 1998, 'The impact of globalisation on women's health', *Arrows for Change Newsletter*, May: 1–2 (Asian-Pacific Resource & Research Centre for Women, Kuala Lumpur).

Shekri N., 2001, India, in W. Wieners, ed., *Global health care markets*. A comprehensive guide to regions, trends, and opportunities shaping the international health arena (San Francisco: Jossey-Bass).

Sidel V.W., H.W. Cohen and R.M. Gould, 2001, 'Good intentions and the road to bioterrorism preparedness', *American Journal of Public Health*, 91: 716–18.

Siegel M., 2001, 'Counteracting tobacco motor sports sponsorship as a promotional tool: is the tobacco settlement enough?', *American Journal of Public Health*, 9(7): 1100–6.

Simmons D. and J.A. Voyle, 1996, 'Psychological and behavioural aspects of NIDDM among Pacific Islanders in South Auckland', *Pacific Health Dialog*, III(1): 100–6.

Simms C., M. Rowson and S. Peattie, 2001, 'The bitterest pill of all: The collapse of Africa's health system', Medact, Save the Children Fund Briefing Report (London: SCF UK).

Simons J., 1989, 'Cultural dimensions of the mother's contribution to child survival', in J.C. Caldwell and G. Santow, eds, *Selected Readings in the Cultural, Social and Behavioural Determinants of Health*, Health Transition Series No. 1 (Canberra: Health Transition Centre, Australian National University).

Simpson D., 2001, 'News Analysis: BAT's internet marketing plan', *Tobacco Control*, 10: 92.

Singer M., 1994, 'AIDS and the health crisis of the US urban poor: The perspective of critical medical sociology', *Social Science and Medicine*, 39(7): 931–48.

Sklair L., 2001, *The Transnational Capitalist Class* (Oxford: Blackwell).

Slade J., L.A. Bero, P. Hanauer, E.D. Barnes and S.A. Glantz, 1995, 'Nicotine and addiction: the Brown and Williamson documents', *Journal of the American Medical Association*, 274: 225–33.

Smith P., 1998, 'The BSE and nvCJD crisis: Implications for public health policy', Lecture presented at the London School of Hygiene and Tropical Medicine, London, 10 December.

Smith J. and R. Pagnucco, 1998, 'Globalizing human rights: The work of trans-national human rights NGOs', *Human Rights Quarterly*, 20(2): 379–412.

Speck R.S., 1993, 'Cholera', in K.F. Kiple, ed., *The Cambridge World History of Human Diseases* (Cambridge: Cambridge University Press), p. 647.

Spegal R., 2001, 'Eating themselves to death', *Pacific Magazine*, September.

Starr D., 1997, *Blood, An Epic History of Medicine and Commerce* (New York: Alfred A. Knopf).

Steinglass M., 2001, 'It takes a village healer', *Lingua Franca*, 28–30 April.

Steinhauer J., 2001, 'U.N. redefines AIDS as political issue and peril to poor', *New York Times*, 28 June.

Stewart F., 1992, 'Can adjustment programmes incorporate the interests of women?', in: H. Afshar and C. Dennis, eds, *Women and Adjustment Policies in the Third World* (London: Macmillan).

Stocker K., H. Waitzkin and C. Iriart, 1999, 'The exportation of managed care to Latin America', *New England Journal of Medicine*, 340: 1131–6.

Stockholm Declaration, 1972, 'Declaration of the United Nations Conference on the Human Environment', *International Legal Materials*, 11: 1416–21.

Stott R., 1999, 'The World Bank: friend or foe?', *British Medical Journal*, 318: 822–3.

Sunley E., A. Yurekli and F. Chaloupka, 2000, 'The design, administration, and potential revenue of tobacco excises', in P. Jha and F. Chaloupka, eds, *Tobacco Control in Developing Countries* (Oxford: Oxford University Press), pp. 409–26.

Tarabusi C.C. and G. Vickery, 1998, 'Globalisation in the pharmaceutical industry, Part 1', *International Journal of Health Services*, 28(1): 67–105.

Tarantola D., 1996, 'Grande et petite histoire des programmes sida', *Le journal du sida*, 86(7): 109–16.

Tarimo E. and E.G. Webster, 1994, *Primary Health Care Concepts and Challenges in a changing world: Alma-Ata revisited*, Current Concerns SHS Paper number 7, WHO/SHS/CC/94.2 (Geneva: World Health Organization).

Tauxe R., E. Mintz and R. Quick, 1995, 'Epidemic cholera in the new world: translating field epidemiology into new prevention strategies', *Emerging Infectious Diseases*, 1(4): 141–6.

Taylor Associates International, 1997, *Private hospital Investment Opportunities*, A global study conducted on behalf of the International Finance Corporation (Annapolis, MD: Taylor Associates International).

Taylor A.L. and D.W. Bettcher, 2000, 'WHO Framework Convention on Tobacco Control: A global "good" for public health', *Bulletin of the World Health Organization*, 78: 920–9.

Taylor A.L., F. Chaloupka, E. Gundon and M. Corbett, 2000, 'The impact of trade liberalization on tobacco consumption', in P. Jha and F. Chaloupka, eds, *Tobacco Control in Developing Countries* (Oxford: Oxford University Press), pp. 343–64.

Taylor S., 2000, 'Drugs firms fight Lamy', *European Voice*, 14 March http://www.european voice.com/this week/index. html.

Tesh S.N., 1988, *Hidden Arguments, Political ideology and disease prevention policy* (New Brunswick and London: Rutgers).

Thaman R.R., 1988, 'Health and nutrition in the Pacific islands: development or underdevelopment', *GeoJournal*, 16: 211–27.

Thomas C., 1987, *In Search of Security: The Third World in International Relations* (Brighton: Wheatsheaf).

Thomas C., 1989, 'On the health of International Relations and the international relations of health', *Review of International Studies*, 15(3): 273–80.

Thomas C., ed., 2000, *Global Governance, Development and Human Security* (London: Pluto).

Thomas S. and Quinn S., 1991, 'The Tuskegee Syphilis Study, 1932–1972: Implications for HIV education and AIDS risk education programs in the black community', *American Journal of Public Health*, 81: 1498–504.

Thornton R., 1990, 'Notes on BAT Group and Associates European Meeting on Technical Issues, Oslo, July 4–5 1990', Guildford Depository, Bates No.: 401350834-837.

Times, The, 2000, 'Chinese cigarettes to be sold in UK', 18 December.

Timmons N., 1995, 'Inequality is worst for 50 years', *The Independent*, 10 February.

Tobacco News, 1994a, 'The Cigarette Industry – Japan', February, Guildford Depository, Bates No.: 500625508.

Tobacco News, 1994b, 'Feature: Smoke Signals', *Tobacco News*, The Tobacco Institute of India, 17–23 July, Bates No.: 301710302.

Tow W., and R. Trood, 2000, 'Linkages between traditional security and human security', in W. Tow, R. Thakur and I. Hyuen, eds, *Asia's Emerging Regional Order* (Tokyo: United Nations University) pp. 13–32.

Treaty of Nice, 2001, 'Amending the Treaty on European Union, The Treaties Establishing the European Communities and Certain Related Acts', *Official Journal of the European Communities*, 10 March, C 80/1.

Tshabalal-Msimang, 2000, 'Cheaper AIDS drugs for South Africa? Minister tells of progress', Health Systems Trust, 24 October 2000, at http://hst.org.za/view. php3?id=20001004.

Tuchman B.W., 1984, *The March of Folly* (London: Michael Joseph Ltd).

Turshen M., 1989, *The Politics of Public Health* (London: Zed Books).

Tysome T., 2000, 'Charities fume over tobacco funding', *Times Higher Education Supplement*, 8 December.

Ugalde A. and J. Jackson 1995, 'The World Bank and international health policy: A critical review', *Journal of International Development*, 7(3): 525–42.

UN, 1945, 'Charter of the United Nations', in I. Brownlie, ed., 1995, *Basic Documents in International Law*, 4th edn (Oxford: Clarendon Press): 1–35.

UN, 1962, General Assembly Resolution on Permanent Sovereignty Over Natural Resources, G.A. Res. 1803 (XVII), 14 December.

UN, 1992, 'United Nations Convention on Biodiversity', *International Legal Materials*, 31: 818–41.

UN, 1994, 'United Nations Convention on Desertification', *International Legal Materials*, 33: 1328–82.

UN, 2001, *The impact of the Agreement on Trade-Related Aspects of Intellectual Property Rights on human rights*, Report of the High Commissioner, E/CN.4/Sub.2/2001/13, 27 June, 53rd Session of the Sub-Commission on the Promotion and Protection of Human Rights.

UNAIDS, 2000a, 'Debt for AIDS Activity', Update February–May, Geneva.

UNAIDS, 2000b, *World AIDS Report* (Geneva: World Health Organization).

UNAIDS Expert Group, 2001, 'AIDS: The Time to Act', UNAIDS Press Release, 10 May at http://www.unaids.org/whatsnew/press/eng/pressarc01/montpelerin_100501reldoc.html.

UN Commission on Human Rights, 2000, Report of the 56th Session on Human Rights, Item 10 (Geneva: United Nations).

UN Commission on Human Rights, 2001, *Intellectual Property Rights and Human Rights*, Report of the Secretary General, E/CN.4/Sub.2/2001/12, 14 June.

UNDP, 1993, *Human Development Report 1993* (New York: United Nations Development Programme/Oxford University Press).

UNDP, 1994, *Pacific Human Development Report* (Suva: United Nations Development Programme).

UNDP, 1997, *Human Development Report 1997* (New York: United Nations Development Programme/Oxford University Press).

UNDP, 1998, *Human Development Report* (New York: United Nations Development Programme).

UNDP, 1999, *Human Development Report: Globalisation with a Human Face* (New York: Oxford University Press).

UNDP, 2001, *Human Development Report 2001. Making new technologies work for human development* (New York: Oxford University Press) http://www.undp.org/hdr2001.

UNECE, 1979, 'Geneva Convention on Long-Range Transboundary Air Pollution', *International Legal Materials*, 18: 1442–50.

UNECE, 1992, 'Convention on the Protection and Use of Transboundary Watercourses and International Lakes', *International Legal Materials*, 31: 1312–29.

UNECE, 2001, 'Environment' at http://www.unece.org/leginstr/cover.html.

UNICEF, 1994, *State of the World's Children* (Oxford: Oxford University Press).

UNICEF, 1996, *State of the World's Children* (Oxford: Oxford University Press).

US, 2000, *The Global Infectious Disease Threat and Its Implications for the United States*, Report of the US National Intelligence Council, Washington DC, January.

USAID, undated, 'Building partnerships to stop AIDS', mimeo, US Agency for International Development, Washington DC.

US Committee on International Science, Engineering, and Technology Policy (CISET), 1995, *Global Microbial Threats in the 1990s* (Washington DC: Working Group on Emerging and Re-emerging Infectious Diseases).

US Department of Health and Human Services, 1988, 'Tobacco Use Among U.S. Racial/Ethnic Minority Groups – African Americans, American Indians and Alaska

Natives, Asian Americans and Pacific Islanders, and Hispanics: A Report of the Surgeon General', (Atlanta, US: US Department of Health and Human Services) http://www.cdc.gov/tobacco/sgr/sgr_1998/sgr-min-pdf/sgr-all.pdf (accessed 13 August 2001).

US General Accounting Office, 1992, *Foreign Assistance; Combating HIV/AIDS in Developing Countries* (National Security and International Affairs Division).

van Bergen J., 1996, 'Epidemiology and Health Policy – A world of difference? A case-study of a cholera outbreak in Kaputa Distict, Zambia', *Social Science and Medicine*, 43(1): 93–9.

van der Vliet V., 2001, 'AIDS: losing "the new struggle"?', *Daedalus*, Winter: 151–84.

Vateesatokit P., 1997, 'Tobacco control in Thailand', *Mahidol Journal*, 4(2): 73–82.

Vateesatokit P., B. Hughes and B. Rittiphakdee, 2000, 'Thailand: winning battles, but the war's far from over', *Tobacco Control*, 9: 122–7.

Vaughan P., S. Mogedal, S.E. Kruse, K. Lee, G. Walt and K. de Wilde, 1995, *Co-operation for Health Development: Extrabudgetary Funds and the World Health Organisation* (Oslo: Governments of Australia, Norway, and the United Kingdom).

Velasquez G. and P. Boulet, 1997, *Globalisation and Access to drugs*, Implications of the WTO/TRIPS Agreement, Health Economics and Drugs, DAP Series No. 7 (Geneva: World Health Organization).

Velasquez G. and P. Boulet, 1999, *Globalisation and Access to drugs*, Implications of the WTO/TRIPS Agreement, Health Economics and Drugs, Revised, DAP Series No. 7 (Geneva: World Health Organization).

Vienna Convention, 1985, 'Vienna Convention for the Protection of the Ozone Layer', *International Legal Materials*, 26: 1529–40.

Volmink J. and P. Garner, 1997, 'Systematic review of randomised controlled trials of strategies to promote adherence to tuberculosis treatment', *British Medical Journal*, 315(7120): 1403–6.

Walker B., 1994, 'A new agenda for AIDS research', *British Medical Journal*, 311: 1448–9.

Wallace R. and D. Wallace, 1994, 'US apartheid and the spread of AIDS to the suburbs: A multi-city analysis of the political economy of spatial epidemic threshold', *Social Science and Medicine*, 41(3): 333–45.

Walsh J.A. and K.S. Warren, 1979, 'Selective primary health care: an interim strategy for disease control in developing countries', *New England Journal of Medicine*, 301(18): 967–74.

Walt G. and J.P. Vaughan, 1981, *An Introduction to the Primary Health Care Approach in Developing Countries* (London: Ross Institute of Tropical Hygiene).

Warden J., 1998, 'Delay in plasma replacement', *British Medical Journal*, 317: 1409.

Waterman P., 1998, *Globalisation, Social Movements and the New Internationalism* (London: Cassell).

Watts S., 1997, 'Cholera and civilization: Great Britain and India, 1817 to 1920', in S. Watts, *Epidemics and History, Disease, Power and Imperialism* (New Haven: Yale University Press), pp. 167–212.

Weber J.T., W.C. Levine, D.P. Hopkins and R.V. Tauxe, 1994, 'Cholera in the United States, 1965–1991', *Archives of Internal Medicine*, 154: 551–6.

Wellman C., 2000, 'Solidarity, the individual and human rights', *Human Rights Quarterly*, 22(3): 639–57.

Werner D. and D. Sanders, 1997, *Questioning the Solution: the Politics of Primary Health* (Palo Alto, CA: Health Wrights).

Wetter D.C., W.E. Daniel and C.D. Treser, 2001, 'Hospital preparedness for victims of chemical or biological terrorism', *American Journal of Public Health*, 91: 710–16.

Whelan D., 1999, *'Gender and AIDS: Taking stock of research and programmes* (Geneva: WHO UNAIDS).

Whidden R., 1990, India – Attached Advertisement (Golden Tobacco Company), 18 December, Bates nos: 2023240555–0556, www.legacy.library.ucsf.edu.

Whitaker M., 1997, 'Mean streets', *New Internationalist*, January/February.

Whiteside A. and T. Barnett, 1998, 'AIDS in Africa: socio-economic determinants and development impact', *IAS Newsletter*, 9, March: 8–11.

WHO, 1992a, 'AIDS: A challenge to science and industry', *World AIDS Day Features*, No. 4.

WHO, 1992b, 'EPI for the 1990s', unpublished document WHO/EPI/GEN/92.2, (Geneva: World Health Organisation) cited in Tarimo and Webster, 1994.

WHO, 1993, *Guidelines for Cholera Control* (Geneva: World Health Organization).

WHO, 1994, *The HIV/AIDS Pandemic; 1994 Overview*, Global Programme on AIDS (Geneva: World Health Organization).

WHO, 1995, *The World Health Report 1995: Bridging the Gap* (Geneva: World Health Organization).

WHO, 1996a, 'Bovine Spongiform Encephalopathy (BSE)', Fact Sheet No. 113, November.

WHO, 1996b, 'International experts propose measures to limit spread of BSE and reduce possible human risks from disease', Press Release, WHO/28, 15 April, Geneva.

WHO, 1996c, 'Report of a WHO consultation on public health issues related to human and animal transmissible spongiform encephalopathies', 2–3 April, Geneva.

WHO, 1996d, 'Scientific consultation on human and animal spongiform encephalopathies', Press Release, WHO/38, 17 May, Geneva.

WHO, 1996e, *The World Health Report 1996: Fighting Disease, Fostering Development* (Geneva: World Health Organization).

WHO, 1997a, 'Typhoid Fever Fact Sheet', at http://www.who.int/inf-fs/en/fact149.html.

WHO, 1997b, 'The Medical Impact of the Use of Antimicrobials in Food Animals: Report of a WHO Meeting' (Geneva: World Health Organization).

WHO, 1998a, 'Emerging and re-emerging infectious diseases', WHO Fact Sheet No. 97, July, Geneva.

WHO, 1998b, *World Health Report 1998 – Life in the 21st Century: A Vision for All* (Geneva: World Health Organization).

WHO, 1998c, *Health for All in the Twenty-first Century*, Document A51/5 (Geneva: World Health Organization).

WHO, 1998d, *Global Cholera Update* (Geneva: World Health Organization).

WHO, 1998e, *Health for All: Policy for the 21st Century*, WHO Document WHA51/5, May (Geneva: World Health Organization).

WHO, 1998f, 'Malaria Fact Sheet', at http://www.who.int/inf-fs/en/fact094.html.

WHO, 1998g, 'Revised Drug strategy, WHO's work on pharmaceuticals and essential drugs', EB/RDS/RC/1, 10 July (Geneva: World Health Organization).

WHO, 1998h, 'Resolution of the Executive Board', EB 101.R24 (Geneva: World Health Organization).

WHO, 1999a, *World Health Report 1999* (Geneva: WHO), http://www.who.int/whr/ 1999/en/report.html (accessed 8 August 2001).

WHO, 1999b, *World Health Report on Infectious Diseases 1999: Removing Obstacles to Healthy Development* (Geneva: World Health Organization).

WHO, 1999c, 'Resolution of the Executive Board', EB 103.R1 (Geneva: World Health Organization).

WHO, 2000a, 'Food safety and foodborne illness', Factsheet No. 237, September http://www.who.int/inf_fs/en/fact237.html.

WHO, 2000b, 'Drug Resistance Threatens to Reverse Medical Progress', Press Release WHO/41, 12 June.

WHO, 2000c, *World Health Report on Infectious Diseases 2000: Overcoming Antimicrobial Resistance* (Geneva: World Health Organization).

WHO, 2000d, 'Tuberculosis Fact Sheet', at http://www.who.int/inf-fs/en/fact104.html.

WHO, 2000e, 'WHO Global Principles for the Containment of Antimicrobial Resistance in Animals Intended for Food: Report of a WHO Consultation' (Geneva: World Health Organization).

WHO, 2000f, 'Revision of the International Health Regulations: Progress Report, July 2000', *Weekly Epidemiological Record*, 75: 234–6.

WHO, 2001a, 'Cholera Fact Sheet', at http://www.who.int/vaccines/intermediate/cholera.html.

WHO, 2001b, 'Anti-Infective Drug Resistance Surveillance and Containment', at http://www.who.int/emc/amr.html.

WHO, 2001c, 'Anti-Infective Drug Resistance, The Role of WHO', at http://www.who.int/emc/amr.html.

WHO, 2001d, 'Revision of the International Health Regulations: Progress Report, February 2001', *Weekly Epidemiological Record*, 76: 61–3.

WHO, 2001e, 'Framework Convention on Tobacco Control', at http://tobacco.who.int/en/fctc/index.html.

WHO and UNICEF, 1978, *Report of the International Conference on Primary Health Care, Alma-Ata, USSR, 6–12 September* (Geneva: World Health Organization).

WHO/Health & Welfare Canada/Canadian Public Health Association, 1986, Ottawa Charter for Health Promotion. From the First International Conference on Health Promotion, Ottawa, 21 November, WHO/HPR/HEP/95.1 (Geneva: World Health Organization).

Wilkinson R., 1992, 'Income distribution and life expectancy', *British Medical Journal*, 304: 165–8.

Wilkinson R., 1996, *Unhealthy societies, The afflictions of inequality* (London: Routledge).

Williams F., 2000, 'UN calls for $3bn to combat AIDS in Africa', *Financial Times*, 29 November, p. 18.

Wilson M., 1996, 'Travel and the emergence of infectious diseases', *Emerging Infectious Diseases* 1(2): 39–46.

Wilson S., 2001, 'Feeding the whole world safely', International Food Policy Research Institute, News & Views, August http://www.ifpri.org/2020/newslet.

Winokur M., 1994, 'IARC update 1', 17 January http://www.pmdocs.com, Bates No.: 2025493459-3460.

Wintour P. and K. Maguire, 2000, 'The funding scandal that just won't go away', *Guardian*, 20 September.

Woelk G., S. Mtisi and J.P. Vaughan, 2000, 'Political economy of tobacco control in Zimbabwe', in J.P. Vaughan, J. Collin and K. Lee, eds, *Case Study Report: Global Analysis Project on the Political Economy of Tobacco Control in Low- and Middle-Income Countries* (London: London School of Hygiene and Tropical Medicine).

Wolfgang H., W.H. Reinicke and F. Deng, 2000, *Critical Choices: The United Nations, Networks, and the Future of Global Governance* (Ottawa: International Development Research Center).

Woodward D., N. Drager, R. Beaglehole and D. Lipson, 2001, 'Globalization and health: a framework for analysis and action', *Bulletin of the World Health Organization*, 79(9): 875–81.

Workers Inquiry into BSE/CJD, 1998, *Human BSE, Anatomy of a health disaster* (London: IW Books).

World Bank, 1987, *Financing Health Services in Developing Countries: An Agenda for Reform* (Washington, DC: World Bank).

World Bank, 1993, *World Development Report: Investing in Health* (New York: Oxford University Press).

World Bank, 1994, *World Bank HIV/AIDS Activities*, Population, Health and Nutrition Department (Washington, DC: World Bank).

World Bank, 2000, 'WHO Calls for Third World Medical Research', World Bank Development News, 11 October.

Worster A., 1980, 'Telex, RJ Reynolds Tobacco Company, July 10', Bates No.: 500882050 at 2051, http://www.rjrtdocs.com, cited: Committee of Experts (2000).

Wright, 1999, 'Does US Trade Policy Keep Aids Drugs Out of Reach?', 26 July, at http://hivinsite.ucsf.edu/social/spotlight/2098.4374full.html.

WTO, 1994a, *Agreement on Trade-related Aspects of Intellectual Property Rights (TRIPS Agreement)*, Marrakesh, 15 April.

WTO, 1994b, *Agreement on Application of Sanitary and Phytosanitary Measures* (SPS Agreement), Marrakesh, 15 April.

WTO, 1994c, *General Agreement on Trade in Services (GATS Agreement)*, Marrakesh, 15 April.

WTO, 1994d, Understanding on Rules and Procedures Governing the Settlement of Disputes, Marrakesh, 15 April.

WTO, 1995, *Trading for the future* (Geneva: World Trade Organization).

WTO, 1997, EC measures concerning meat and meat products (hormones) complaint by the United States, Dispute settlement, Panel Report (Geneva: World Trade Organization).

WTO, 1998a, EC Measures concerning meat and meat products (hormones), WT/DS26/AB/R. Report of the Appellate Body, 16 January (Geneva: World Trade Organization).

WTO, 1998b, EC Measures concerning meat and meat products (hormones), WT/DS26/15, WT/DS48/13, Arbitration under Article 21.3.(c) of the Understanding on Rules and Procedures Governing the Settlement of Disputes (Geneva: World Trade Organization).

WTO, 1998c, *Council for Trade in Services. Health and Social Services*, S/C/W/50, 18 September (Geneva: World Trade Organization).

WTO, 2000a, WC Measures affecting asbestos and asbestos-containing products, Report of the Panel, WT/DS135/R, 18 September (Geneva: World Trade Organization).

WTO, 2000b, Council of Trade in Services, Communication from the European Communities, GATS 2000, Environmental services, WTO/CSS/W/38 22 December.

WTO, 2000c, Council of Trade in Services, Communication from the European Communities, GATS 2000, Transport services, WTO/CSS/W/41 22 December.

WTO, 2001a, Council of Trade in Services, Communication from the European Communities, GATS 2000, Postal and courier services, WTO/CSS/W/61 23 March.

WTO, 2001b, Council of Trade in Services, Communication from the European Communities, GATS 2000, Energy services, WTO/CSS/W/60 23 March.

WTO, 2001c, EC Measures affecting asbestos and asbestos containing products, Report of the Apellate Body, 12 March, WT/DS/AB/R (Geneva: World Trade Organization).

Yach D. and D. Bettcher, 1999, 'Globalization of tobacco marketing, research and industry influence: Perspectives, trends and impacts on human welfare', *Development*, 42(4): 25–30.

Yach D. and D. Bettcher, 2000, 'Globalisation of tobacco industry influence and new global responses', *Tobacco Control*, 9: 206–16.

Young O.R., 1997, *Global Governance: Drawing insights from environmental experience* (Cambridge, MA: MIT Press).

Zakaria F., 2000, 'Globalization grows up and gets political', Opinion in *The New York Times on the web*, 31 December http://www.nytimes.com/2000/12/31/opinion/31ZAKA.html.

Zwi A., 1981, 'Cholera in South Africa', *South African Outlook*, 11: 172–7.

Zwi A. and A. Cabral, 1991, 'Identifying "high risk situations" for preventing AIDS', *British Medical Journal*, 303: 1527–9.

Index